C000157043

Influencing Adversary States

Quelling Perfect Storms

PAUL K. DAVIS, ANGELA O'MAHONY, CHRISTIAN CURRIDEN, JONATHAN LAMB

Prepared for the Office of the Secretary of Defense for Policy
Approved for public release; distribution unlimited

RAND NATIONAL DEFENSE RESEARCH INSTITUTE

For more information on this publication, visit www.rand.org/t/RRA161-1

Library of Congress Cataloging-in-Publication Data is available for this publication.
ISBN: 978-1-9774-0652-1

Published by the RAND Corporation, Santa Monica, Calif.
© 2021 RAND Corporation
RAND® is a registered trademark.

Cover: cherylvb/Adobe Stock; fotokostic/Getty Images.
Cover design: Katherine Wu

Support RAND
Make a tax-deductible charitable contribution at
www.rand.org/giving/contribute

www.rand.org

Preface

This report describes experimental concepts and methods for assisting in the formulation of strategy to deter aggression—or escalation thereafter—by major state adversaries and competitors. It uses a hypothetical crisis with China to illustrate necessary contextual information. Comments or questions should be addressed to Paul K. Davis and Angela O'Mahony at pdavis@rand.org and aomahon1@rand.org.

The research reported here was completed in September 2020 and underwent security review with the sponsor and the Defense Office of Prepublication and Security Review before public release.

This research was sponsored by the Strategy and Force Development office within the Office of the Secretary of Defense for Policy (OSDP/SFD) and conducted within the International Security and Defense Policy (ISDP) Center of the RAND National Security Research Division (NSRD), which operates the National Defense Research Institute (NDRI), a federally funded research and development center sponsored by the Office of the Secretary of Defense, the Joint Staff, the Unified Combatant Commands, the Navy, the Marine Corps, the defense agencies, and the defense intelligence enterprise.

For more information on the RAND ISDP, see www.rand.org /nsrd/isdp or contact the director (contact information is provided on the webpage).

Contents

Figures

Tables

Summary

In this report, we describe an experimental thinking-Red approach to analysis, wargaming, and other exercises to inform strategies for avoiding enemy aggression or escalation in a crisis. Such an approach anticipates (with alternative Red models) how Red may be reasoning; it then seeks ways to influence Red's reasoning in ways that we (Blue) regard as favorable. An influence strategy might involve a mix of deterrence by threat of punishment, deterrence by denial, dissuasion by various means, reassurances, and incentives. Deterrent threats alone will seldom constitute effective strategy and, depending on the adversary's motivations and perceptions, could even be counterproductive. A successful strategy will also often require artful orchestration of political, military, and economic instruments of power.

Background

In the first part of the report, we recount a long history of high-level crisis decisionmaking that is replete with examples of blunders that could have been avoided. We then relate these blunders to well-studied psychological shortcomings in human reasoning. These examples present a set of challenges that decision-aiding should try to mitigate. We then describe an approach to mitigation that could draw on analysis and—of most interest in this report—such activities as wargames and other human exercises. Our approach focuses primarily on thinking more carefully about the adversary, referred to as *Red*—but recognizing that it is notoriously difficult to do so well.

We urge constructing alternative models of the adversary to open minds, overcome the common tyranny of the best estimate, and better fashion hedged influence strategies. We seek to avoid both the error of mirror-imaging (imagining that the adversary perceives and reasons as we do) and seeing the adversary as a stereotype (e.g., one that only perceives and reasons in a rigid, malevolent, and unintelligent way focused on aggression and war-winning). Our approach urges assessing the adversary's possible motivations for aggression and the adversary's reasoning about the acceptability and attractiveness of options for that aggression. We do this not in the context of peacetime circumstances, in which competition may be strong but the potential for war is low; instead, we do this for circumstances in which conflict would be most plausible. Such circumstances might include the adversary having fears and a sense of desperation, aggressive ambition, or both. We refer to this as *anticipating Perfect Storm* circumstances. This means that we are not expecting or planning for the worst-case scenario but instead are attempting to aid planning to avoid the worst case.

Approach

Figure S.1 shows an overview of the approach, displaying the analytical steps to be taken in background staff work, wargames, and other exercises, with the intent of using the fruits from these steps to inform strategy formulation. The decision-aiding itself could include analysis papers written by staff, arguments by such principals as the Secretary of Defense, or aids for policymakers and their staff when they participate in exercises or wargames.

As indicated by the stacked icons (⬚) in Figure S.1, we begin by recognizing and articulating alternative models of the adversary (*Red*). These alternatives should go well beyond the day's best estimate of Red or the image of Red held by dominant personalities. They should be taken seriously and used analytically. The less-popular models should not be treated with polite derision, as often happens when someone is asked to serve as the devil's advocate. Nothing can prevent decisionmakers from ignoring analysis and acting on their personal

Figure S.1
Overview of Approach

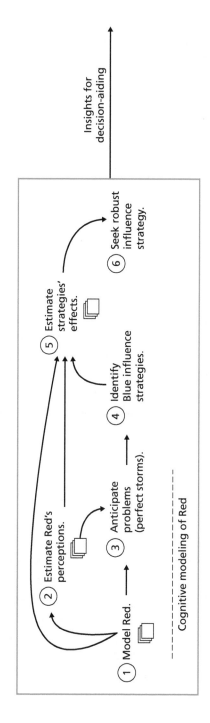

1 Model Red.

2 Estimate Red's perceptions.

3 Anticipate problems (perfect storms).

4 Identify Blue influence strategies.

5 Estimate strategies' effects.

6 Seek robust influence strategy.

Cognitive modeling of Red

Insights for decision-aiding

judgments alone, but a doctrine for decision-aiding should confront the uncertainties and disagreements respectfully and dispassionately. It should also compensate for the problem that intuition, after two decades of countering terrorists and violent extremists, may be inappropriate for dealing with great-power competitors. The next step in the approach is, for each of the alternative Reds, to characterize likely and possible Red perceptions. The third step is to ask about the circumstances that could lead to perfect storms—instances when Red might be motivated to act aggressively. Although not shown in the diagram, we consider both a style of Red reasoning that is heuristic and possibly impulsive and a mode of Red reasoning that would be intendedly deliberate and rational but that would also be subject to misperceptions, miscalculations, and cognitive biases.

Having characterized how Red might reason in dangerous circumstances, the approach then creates a menu of Blue influence actions, from which alternative influence strategies could be constructed. These actions would be intended to affect Red's motivations, perceptions, and calculations. The alternative influence strategies are then compared for their potential effect on the alternative models of Red. An ideal strategy would be effective for all of the alternative Reds. Such a strategy would typically be adaptive, so that the balance among Blue actions would change in the course of events. This contrasts with a strategy that bought so fully into the best-estimate assessment as to preclude adaptation. Such an inflexible strategy might, for example, count on Red being deterred or, in the converse, seeing Red as intent on war.

This approach incorporates many features familiar to senior diplomats, warriors, and presidents. Any novelty in the approach is its effort to enforce a *doctrine of decision-aiding under uncertainty*, which would hedge bets rather than lead to a strategy that was "all-in" with respect to a particular interpretation of the adversary. A frequent result of the application of this doctrine might be a hybrid strategy that takes active steps to deter, prepare for conflict, and address potentially resolvable tensions despite such steps possibly being criticized as weak or naive. The intent is to help decisionmakers find such hybrids despite pressures to act decisively (but perhaps unwisely). The report discusses many historical instances in which national leaders did not have the well-hedged strategies that we have in mind. This was not because no one thought of the uncertainties, but because the social, analytical, and managerial aspects of the decisionmaking system did not encourage hedging. For example, the European powers fell into World War I despite the potential consequences, the United States banked on deterring Japan from attacking Pearl Harbor, top-level advisers urged President John F. Kennedy to invade Cuba even though we now know that doing so might have led to calamity, Israel delayed reacting to warning of Egypt's attack in 1973 because it was confident that Egypt would be deterred, and, closer to home, the United States did not prepare seriously for a counterinsurgency phase after the anticipated successful invasion of Iraq and overthrow of Saddam Hussein.

Practicalities

In the phase of research reported here, we focused primarily on developing experimental methods for strategy-building exercises organized by the Office of the Secretary of Defense. With that in mind, the report includes a chapter describing generically what might be developed in an exercise addressing a particular potential or ongoing crisis, followed by a chapter simulating an exercise involving a hypothetical crisis with China regarding Taiwan. An appendix collects templates and other materials to support future live or virtual exercises with people. We believe that the groundwork has now been laid to support live or virtual exercises to test the approach more carefully and work on improving decision aids for participants.

Although we say little about combined methods in this report, a good approach to decision-aiding should combine modeling, analysis, and human exercises. For broad exploratory analysis, a computational version of the approach is needed—one that confronts many more dimensions of uncertainty than can be treated in a given exercise. We have developed an initial version of such a computational model, described briefly in an appendix. We believe that we have laid the basis for subsequent testing and experimental applications.

Next Steps

We see strategy-informing applications to (1) different potential adversaries, (2) gray-zone conflicts, (3) different levels of crisis or conflict, and (4) anticipation of competitor actions against alliances and partnerships. Each of these applications will require in-depth study of substantive issues and refinement of methods and tools.

Acknowledgments

We thank Thomas Szayna at the RAND Corporation and John Arquilla at the U.S. Naval Postgraduate School for their careful in-depth reviews. We also benefited from ideas suggested by RAND colleagues Peter Wilson, Cortez Cooper, Bonnie Lin, and Derek Eaton.

CHAPTER ONE

Introduction

Motivations for the Study

As discussed in the most recent *National Defense Strategy* (Mattis, 2018, p. 2), the United States has major national security challenges:

> The central challenge to U.S. prosperity and security is the *reemergence of long-term, strategic competition* [emphasis original] by what the National Security Strategy classifies as revisionist powers. It is increasingly clear that China and Russia want to shape a world consistent with their authoritarian model—gaining veto authority over other nations' economic, diplomatic, and security decisions.

The *National Defense Strategy* specifies pointed defense objectives that include deterring adversaries from aggression against vital U.S. interests, maintaining favorable balances of power in several regions, and defending allies from military aggression. These are familiar, traditional defense objectives, but they were seen to require fresh thinking after many years during which the United States was largely preoccupied with terrorism, insurgencies in the Middle East, and other matters distracting attention from challenges posed by China and Russia. Furthermore, fresh thinking is needed because power balances and the range of strategies used by aggressor nations have changed. This report is intended to contribute to that fresh thinking.

Objectives

In this report, we describe and experiment with a new approach to informing the development of strategy to deter aggression against the United States or its allies or to limit escalation in the event that initial deterrence fails. The strategy is intended to be used in strategic planning by the U.S. Department of Defense and possibly for crisis decision-aiding. It uses plausible but purely notional crisis situations with China to bring to bear illustrative but notional context and detail. We believe that the new approach could help avoid errors seen in historical examples and aid in the development of integrated, adaptive whole-of-government strategies.

Approach

Our approach in this research is what we call uncertainty-sensitive cognitive modeling. The tenets of the approach are not new. What is perhaps more novel is that we are embedding the tenets within an analytic structure to aid studies, gaming, simulation, and decisionmaking. We use *cognitive modeling* to mean modeling that represents how, *in effect*, decisionmakers reason about their strategic decisions. To be sure, what goes on in the minds of humans is often messy, chaotic, and inconsistent—with markedly contradictory beliefs in play (Jervis, 2006a). Nonetheless, as people attempt to reason sensibly, certain factors come to loom large, and a nominal structure emerges. This is what we seek to describe, something akin to a leader summarizing to his top deputies as follows:

> Well, to summarize, we worry primarily about ___, ___, and ___, but we also pay attention to ___ and ___. The options have different strengths and weaknesses, but we are convinced after spirited and sometimes anguished debate that we must assure ___ and ___, and we must "balance" the other considerations. All things considered, it seems that Option B is probably best—although we may have to revisit this later depending on initial results. Option B should get good results

if matters develop as expected, but it is also hedged: It will probably do fairly well even if matters develop more unfavorably and it will allow us to do exceptionally well if we are lucky.

We had almost agreed on a different course of action last week, but debate revealed that we were accepting uncritically one key assumption—i.e., that ___. Recognizing this, we took the time to double-check with the field units who are closest to the problem, looked at historical experience, and tested against our most detailed models. We made some changes as a result. We also came to realize last week that we had been ignoring the possibility of side effects that we would find appalling, such as that ___. That is, we had not been recognizing the threat to some of our strongest values, of which we had been insufficiently conscious. When we realized that, we again did some serious rethinking. As a result, Option B now ends up looking to be our best choice. Do we agree?

A military commander's summary to his or her subordinate officers would be different and shorter, since he or she would be announcing a decision, but the same idea applies.

This imagined reasoning is strategic but human in character. It incorporates uncertainty-sensitive analysis by uncovering important assumptions, evaluating the options under different assumptions, and seeking an option that should do reasonably well across the range of plausible assumptions. The reasoning refers to going into detail to check some matters—multi-resolution analysis done as necessary background work. In contrast with an academic exercise in decision analysis that inputs utilities (as though those are known in advance and stable), the summary notes that the group has been *discovering* some of its values along the way. This imagery, then, is the motivation for our general approach, which draws on substantial past research.[1]

---

[1] The earlier research refers to uncertainty-sensitive planning (Davis, 2003b), assumptions-based planning (Dewar, 2002), exploratory analysis (Davis, 2003a), and Robust Decision Making (Lempert, Popper, and Bankes, 2003; Marchau, Walker, Bloemen, and Popper, 2019).

Background

Deterrence

Humans have been attempting to deter individuals, tribes, or countries for millennia. Statesmen and generals have forever sought to formulate strategies to accomplish this. History, however, admonishes caution and humility. Despite leaders' best efforts, historical failures of deterrence are legion (see Chapter Two for more on this). Problems with definitions also arise, so we collect our definitions of deterrence and related concepts are in Table 1.1.

Failures of deterrence today are not primarily the failures of theory. Scholars in political science, psychology, economics, and decision analysis have studied deterrence and the related subject of coercive diplomacy for decades. The topic is well understood and articulated, including the circumstances in which deterrence fails, as discussed in a recent review (Mazarr, 2018; Mazarr, Chan, Demus, Frederick, Nader, Pezard, Thompson, and Treyger, 2018). Drawing on this recent review and our own past research,[2] the following points loom large. They touch in sequence on understanding the potential aggressor's motivation, five aspects of deterrence, the role of perceptions, and the need to see deterrence as merely one component of influence.[3] Our discussion is deliberately terse because so much prior literature exists. Table 1.1 defines most of the terms that follow.

Motivation is a major factor in decisions. Deterrence often fails because of fear, desperation, and a perceived necessity of acting and not because of aggressive opportunism per se (Lebow, 1983). The motivation may be the result of internal considerations that are not subject to external influence.

---

[2] A critical review of deterrence theory updates for the context of U.S. extended deterrence on behalf of South Korea (Davis, Wilson, Kim, and Park, 2016). Aspects were incorporated in National Research Council of the National Academies, 2014.

[3] See George and Smoke, 1974; George, 2003. See also chapters on deterrence (Levy, 2008) and crisis management (Stein, 2008) in a testimonial volume on Alexander George's seminal contributions to political psychology (Renshon, 2008). George saw deterrence and coercive diplomacy as part of influence (George, 2003, p. 272). See also Schultz, 2001.

Table 1.1
Definitions

Term	Meaning
Influence	Effects on the decisions of another party by, for example, positive inducements, persuasion, dissuasion, deterrence, compellence, and punishment
General deterrence	Deterrence over time in periods of peace. If successful, it will head off crises in which immediate deterrence would be at issue
Deterrence (classic)	Persuading an adversary not to take an action by threatening punishment only if the action is taken but not otherwise (see also "broad deterrence," below)
Dissuasion by denial (often called *deterrence by denial*)	Persuading an adversary not to take an action by having the perceived capability to prevent success adequate to justify the costs
Cumulative deterrence	Quality of deterrence at a given time because of the history of prior successful and failed deterrent actions, crises, and conflicts
Broad deterrence	A combination of deterrence, dissuasion by denial, and cumulative deterrence
Direct deterrence	Deterring an attack on the United States or its immediate interests; more likely to succeed than extended deterrence (see next entry), because the deterrent threat is inherently more credible
Extended deterrence	Convincing an adversary not to take an action against the interests of an ally by the methods of broad deterrence
Dissuasion	Persuading an actor (such as an adversary) not to take a particular action
Compellence	Causing an actor (such as an adversary) to take an action despite its preferences to the contrary by using or threatening to use military, economic, or political power
Coercion	Causing an actor to unwillingly do something by use of force or threat; deterrence and compellence are different kinds of coercion
Assurance	Convincing an ally that there is U.S. commitment to and capability for extended deterrence for the purpose of dissuading the ally from developing its own nuclear arsenal
Reassurance	Reducing fears of potential adversaries about U.S. or ally intentions

Table 1.1—Continued

Term	Meaning
Coercive diplomacy	Diplomacy that may use incentives and other forms of persuasion to supplement the actual, threatened, or implicitly threatened use of force to persuade an adversary to change its behavior; may include reassurances and incentives.

NOTE: These definitions come from Table 2-1 of National Research Council of the National Academies, 2014. The definition of *coercive diplomacy* is inferred from George, 2003, p. 273; it is also discussed in Schultz, 2001.

Classic deterrence by threat of punishment. Threats may deter, but they are often ineffective. The attacker may not believe that the defender will carry out the threats. Furthermore, the defender is often unwilling to make its intentions unequivocal or its response certain.[4] Even if threats are credible, nations, like individuals, do not react well to threats. Indeed, they have their own values, pride, and politics. As a result, threats can be counterproductive.

Extended deterrence (deterring aggression against an ally) is especially difficult—a relatively thin reed on which to depend (Morgan, 1983; Morgan, 2003; Huth, 1988; Davis, Wilson, Kim, and Park, 2016). The punishment threatened in extended deterrence is often much less credible, as recognized famously by France's Charles de Gaulle by the mid-1950s.

General deterrence is more powerful than immediate deterrence. It is far better not to get into crisis than to manage an existing crisis.[5]

Deterrence by denial. Convincing an adversary that any attack would be defeated, or at least that the attacker would not achieve its objectives, is inherently stronger than deterrence by threat of

---

[4] We are less sanguine than Mazarr, Chan, Demus, Frederick, Nader, Pezard, Thompson, and Treyger, 2018, about the feasibility of the United States making its determination clear and compelling. A nation's values and intentions can vary by its leader and its emerging realities. For example, President George H. W. Bush and his cabinet did not know how they would react to an Iraqi invasion of Kuwait before the invasion happened. Even after the invasion, strenuous arguments were made against military action by General Colin Powell, among others (Woodward, 1991, pp. 205–246; Gans, 2019).

[5] For general deterrence, an actor "maintains a broad military capability and issues broad threats of a punitive response to an attack to keep anyone from seriously thinking about attacking" (Morgan, 2003, p. 9).

Figure 1.1
Influencing the Adversary's Choice of Actions

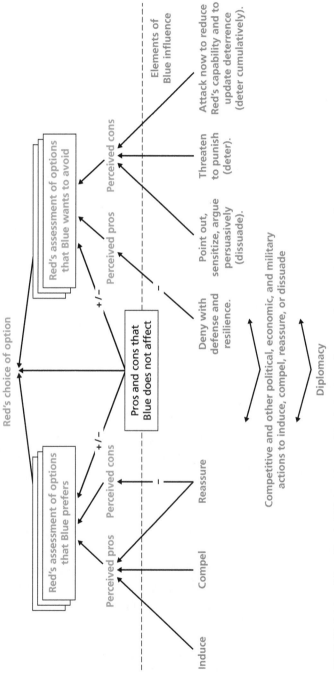

SOURCE: Adapted from Davis, Wilson, Kim, and Park, 2016, and National Research Council of the National Academies, 2014.

punishment. Related military actions, however, could be perceived as threatening and could lead to a counterproductive action-reaction cycle in peacetime or escalation in crisis.[6]

Perceptions. The adversary's perceptions matter in failures of deterrence and influence. These perceptions might or might not correspond to reality. They may or may not be appropriately adjusted in response to political signals.

Escalation

Although this report focuses on crises and the desire to avoid war, the methods it describes can be used to address the problem of escalation (an increase in the level or extent of conflict) if initial deterrence fails. The history of escalation in war suggests that escalation is very dependent on contextual details, including the original path to war, the protagonists' particular experience, and events. Some wars that were expected to escalate did not. Other times, escalation continued despite efforts to avoid it. Results depended in part on how well the antagonists understood each other and how aware they were of the risks involved in escalation (Smoke, 1977).

Putting It Together: Influence Theory

Deterrence should be seen as part of influence. Influence tactics could include reassurances to reduce the adversary's fears, incentives, conditional reciprocity, positive incentives, relationship-building, and both direct and various means of dissuasion. (Dissuasion could be in the form of political, economic, and military actions to convince the adversary that the aggression at issue would not be in its interest [e.g., it might mean lost opportunities for political or economic gain, loss of allies worldwide, or even a pariah status].) The challenge is to persuade the adversary to take an option that we prefer (possibly one that has yet to be recognized) rather than an option of aggression. Figure 1.1 makes the point schematically. The graphic includes terms that are more familiar to some readers than others. Most of the definitions used

---

[6] This refers to what is often described as the *security dilemma*, a term introduced by John Herz in 1950 but discussed most famously by Robert Jervis, 1978.

(see Table 1.1) come from a national academy study (National Research Council of the National Academies, 2014), but we have added the item on coercive diplomacy—or what is sometimes called *forceful persuasion* (George and Simons, 1994). Both deterrence and coercive diplomacy are part of influence theory (George, 2003).[7]

Structure of the Report

We had all of these issues in mind as we developed the approach described in what follows. Despite the often-dismal record of influence failures, our project was based on the belief that it should be possible to formulate and adapt an influence strategy to avoid many of the blunders that might otherwise happen and that qualitative modeling should be helpful in doing so.[8] We describe our approach in the remainder of this report. In Chapter Two, we describe the problems, as we see them, based on historical experience and modern knowledge from psychology about decisionmaking. In Chapter Three, we describe our approach in generic terms. Chapter Four presents our application of the approach notionally to a crisis with China. We then draw some brief conclusions in Chapter Five. The appendixes present our discussion of some mathematical issues, a computational model instantiating a factor-tree model, a Bayesian model for updating the adversary's model of the defender, and some detailed input data for particular purposes.

---

[7] The relationships among deterrence, coercive diplomacy, and other aspects of influence are discussed in a paper written for the 2000s (George, 2003) that builds on more-classic discussion (George and Smoke, 1974, pp. 588–616).

[8] The historical record of blunders motivated President Barack Obama's view that the key to good foreign policy is "don't do stupid shit" (Goldberg, 2016).

CHAPTER TWO

Lessons from History and Psychology

In developing the approach that we describe in Chapter Three, we drew on both history and psychology for insights. We describe those separately in what follows, although many of the historical incidents mentioned were affected by psychological issues.

Historical Blunders

Deterrence has apparently been successful in many historical episodes, although it is often difficult to know whether war was avoided because of deterrence or lack of motivation. In any case, NATO and the Warsaw Pact did not have a World War III, China has not invaded Taiwan, North Korea has not invaded South Korea in the past 70 years, and, some would point out, the United States did not attack North Korea after it tested nuclear weapons and long-range intercontinental ballistic missiles (ICBMs), threatening the United States in 2017. It is useful, however, to think of major historical failures of deterrence, each involving what, in retrospect, were blunders. These are related to why nations suffer surprise attacks (Bracken, Bremer, and Gordon, 2008).

Great-power rivalry leading to World War I. Great Britain, Germany, and the Russian Empire fell into World War I due, in part, to misunderstandings about mutual intentions and under-appreciating risks. Scholars still disagree about the origins of the war and how blame should be allocated, but the major powers effectively concluded that it was better to have a war then than to have the inevitable war later, and that the war was a risk worth taking (Monbauer, 2014). The major powers underestimated the horrific consequences and behaved as though they believed that the war would be short—suppressing their recognition that it might instead be long and terrible.[1] They anticipated an offense-dominated war that created pressures to act quickly to mobilize (Howard, 1984).

U.S. strategy toward Japan before Pearl Harbor. The United States, United Kingdom, and Netherlands imposed economic sanctions on Japan in 1940 and 1941 to deter it from further imperial aggression. They blundered in this coercive diplomacy by not recognizing that it was tantamount to war and would force Japan to respond massively (George and Simons, 1994). President Franklin Roosevelt believed that Japan would be deterred from attacking Pearl Harbor.[2]

Japan's decision to go to war. Japan's leaders also blundered. The blunder was as multifaceted as those leading to World War I. Japan's leadership was well aware that Japan would lose a long war, but it could not bear to give in to the efforts of the Western powers to quash its imperial ambitions. Instead, it felt *compelled* to fight because economic warfare by the United States and Britain was undercutting its imperial ambitions.[3]

---

[1] Leaders can carry along contradictory beliefs and sometimes choose what to believe. Before World War I, for example, leaders were aware that war could be long (Jervis, 2006b, p. 647; Jervis, 2017).

[2] After Japanese aggression in China, Korea, and Indochina, the United States, United Kingdom, and Netherlands imposed trade restrictions, including an oil embargo in 1941. Leaders recognized that this could lead to hostilities. Despite strong warnings from the Navy, President Roosevelt moved the Pacific Fleet to Pearl Harbor as a deterrent measure. Having the fleet at Pearl Harbor, however, provided an attractive and vulnerable target. A review of ADM James O. Richardson's book provides pointers to the contentious literature on culpability for the surprise at Pearl Harbor (Merson, 1988). The review paints an overlapping but different picture than that of the renowned book by Roberta Wohlstetter (1962). One consensus conclusion, in Thomas Schelling's foreword to that book, is that

> It is not that we were caught napping at the time of Pearl Harbor. Rarely has a government been more expectant. We just expected wrong. We were so busy thinking through some "obvious" Japanese moves that we neglected to hedge against the choice that they actually made. (Schelling, 1962)

[3] Eri Hotta describes the view from inside the Japanese government (Hotta, 2013), giving examples of the kinds of decisionmaking errors that we discuss here. Although eager to

Hitler's aggression in Europe. Some leaders in England and France misunderstood Hitler, his intentions, and his reasoning in the wake of the Munich Agreement in 1938. They pursued a policy of appeasement that failed utterly, forever giving the term *appeasement* a bad name.[4] When he was ready, Hitler was not deterred from invading Poland.

The Cuban Missile Crisis (a near disaster). General deterrence failed when the Soviet Union chose to send nuclear missiles to Cuba (seeing it as analogous to NATO having missiles in Turkey). When this was discovered in 1962 and the crisis became acute, senior advisers urged President John F. Kennedy to invade Cuba, believing that the Soviet Union would be deterred from intervention because of U.S. regional and global military superiority. Had the advice been followed, the result might well have been a disaster. The advisers did not know that the Soviets already had nuclear weapons in Cuba and on its submarines, nor did they know that the Soviet commander had—for a time—pre-delegated authority to use nuclear weapons in the event of a U.S. invasion.[5] This, then, was a very near miss.

U.S. escalation in Vietnam. The United States attempted deterrence and coercive diplomacy against North Vietnam, but the North pursued covert aggression nonetheless. The United States then engaged militarily and continued to escalate in an effort to deter further advances by North Vietnam and compel cessation. The Vietnam War was notable for the many strategic and military miscalculations by the United States, including misunderstanding both Ho Chi Minh and the

---

avoid war, Japanese leaders were influenced by reckless militarism and seduced by imagining that—if all went well despite the odds—they could win magnificently.

[4] Classically, *appeasement* is simply attempting to reduce tensions by removing the principal causes of conflict (e.g., making reasonable concessions to eliminate grievances). It can sometimes be pragmatic, but Hitler was unappeasable, and the term has taken on a foul odor (George, 2003, p. 167). The Munich Agreement did, however, buy time at a point of great military weakness (Ripsman and Levy, 2008).

[5] A voluminous literature exists on the Cuban Missile Crisis (e.g., Garthoff, 1988), but alarming discoveries were still being made decades later (Dobbs, 2008). Regarding the advice that Kennedy was receiving, one need not depend on secondary or tertiary sources. Transcripts are available for many top-level discussions (May and Zelikow, 2002), and some of the tapes can be accessed online via the National Security Archives of George Washington University. They are chilling to hear.

Vietnamese people, and the many errors and blatant falsifications in military assessments (McMaster, 1998; Halberstam, 2002). Domestic politics in the United States also had profound consequences. Kennedy, President Lyndon B. Johnson, President Richard Nixon, and Secretary of State Henry Kissinger were strongly motivated to avoid being seen as soft on Communism or responsible for losing the war. This surely biased their judgments about what might be accomplished and helped prolong a failed war.

Egypt's attack of Israel. In 1973, Anwar Sadat's Egypt attacked Israel even though it would have seemed that he would be deterred by Israel's superior military capability. Sadat had his reasons for attacking Israel, however, and did not anticipate a permanent military victory. Instead, his objectives were political, psychological, and strategic.[6] Israel, for its part, blundered by rationalizing away many warning signs until it was too late. As a result, it suffered operational surprise. Interestingly, in this case, the errors depended significantly on personality characteristics within Israeli intelligence that enhanced motivated bias (Bar-Joseph and Kruglanski, 2003).

Argentine actions leading to the Falklands/Malvinas War. Argentina surprised the world in 1982 when it seized the Falkland Islands (Malvinas) by force. Initially, Argentina believed that the UK would not respond militarily, but the UK responded despite its immense distance from the islands and military challenges.[7] Still, Britain failed to deter the seizure. Even when Argentina recognized that Britain would fight, it chose to continue to do so. This was a failure of Britain's coercive diplomacy.

Saddam Hussein's invasion of Kuwait. Saddam Hussein surprised the United States in 1990 by invading Kuwait. The surprise was a

---

[6] A short review of the war's background explains the frustrations behind Sadat's move (Office of the Historian, undated). A well-connected reporter's account from the time is colorful and insightful (Sheehan, 1973). Sadat's own book is fascinating but not to be construed as rigorous history (el-Sadat, 1977).

[7] President Leopoldo Galtieri, head of the junta controlling Argentina, saw the grab of the Falklands (Malvinas) as a way to assuage public concern about rampant inflation and human-rights abuses. He anticipated an easy victory and no military response from the British (Arquilla and Rasmussen, 2001, p. 739).

result of a U.S. misunderstanding of Saddam's strategic reasoning and passions and the United States not recognizing the weakness of its own deterrent in Southwest Asia.[8] Saddam's invasion proved to be a miscalculation, but his decision to invade was rational given his not-unreasonable assumptions about U.S. willingness to fight over Kuwait, his correct recognition that the United States had almost no relevant military forces in the region, and his belief that the Saudis would not welcome U.S. force. The United States blundered in its attempts at coercive diplomacy.

The Soviet invasion of Afghanistan. As a foolish mistake, rather than because of a failure of deterrence by the West, the Soviet Union invaded Afghanistan in 1979. The invasion was due in part to an exaggerated perception of what the Soviet Union saw as growing U.S. influence in Afghanistan (the Soviets feared that Afghanistan would switch loyalties to the West). Some Soviets also feared that losing Afghanistan would be "acknowledging that history can be reversed, setting the stage for the disintegration of the entire empire."[9] Civilian leadership believed that the war would be brief and easy; it also underestimated the military and international consequences of invasion, as well as the subsequent difficulty of extracting itself.[10]

U.S. invasion of Iraq. The United States invaded Iraq in 2003 having seriously underestimated the probability of an extended insurgency and the difficulties of leaving. From the perspective of

---

[8] Early analysis got much of the diagnosis right (Davis and Arquilla, 1991a; Stein, 1992), but much more has been learned from interviews of Saddam's high-level officials and generals and access to many official documents and even tape recordings of Saddam's conversations with generals and officials (Woods and Stout, 2010; Woods, Pease, Stout, Murray, and Lacey, 2006).

[9] This is ascribed to Michael Dobbs in Baker, 2019, apparently referring to Dobbs, 1997, pp. 9, 17, in which Dobbs discusses Leonid Brezhnev feeling responsibility for preserving the Soviet Empire and the principle of historical irreversibility. A primary-source insider account speculates about who convinced the senile Brezhnev to invade and notes that the invasion was soon seen as a disastrous mistake (Chernyaev, 2020).

[10] For primary-source material, see Blanton and Savranskaya, 2019. See also short discussions by Baker, 2019, and Jones, 2019. As another example of how adversaries are misunderstood, Zbegniew Brzezinski, the U.S. national security adviser, wrongly saw the invasion as Russia fulfilling an age-old dream of having direct access to the Indian Ocean.

those who saw the U.S. invasion as illegal, including the Secretary-General of the United Nations (UN) Kofi Annan (Tyler, 2004), the unsanctioned invasion was a violation of the long-standing UN principle of recognizing territorial integrity enshrined as Article 2(4) of the UN Charter.[11] The decision to invade was a miscalculation, but the failure to prepare for the possibility of extended opposition and even insurgency was a blunder.

Russian invasion of Crimea and intervention in eastern Ukraine. After having largely written off any threat from Russia, NATO and other countries worldwide were surprised in 2014 when Vladimir Putin directed Russia's seizure of Crimea and pursued subsequent troublemaking in Ukraine. The blunder was imagining that a broad international code of conduct would deter such moves or that Russia was no longer a threat.

Many factors underlay these instances of flawed decisionmaking, but some generic villains are recognizable from past work on political psychology, deterrence, and adversary modeling. Some of the factors relate to cognitive biases, which are further discussed in the section on psychological errors in Chapter Two, and highlighted here:

1.   Misunderstanding adversaries and their perceptions. This can take several forms. One is the error of mirror-imaging—not recognizing when the adversary (e.g., Hitler, Japan's leadership in the 1930s, or Ho Chi Minh in the 1960s) has a different value structure than we do. Another error is rather the opposite: exaggerating and stereotyping, such as exaggerating the adversary's aggression, hostile intent, and tolerance for human costs (as when critics of détente claimed that the Soviets did not understand or even have a word for deterrence and were instead convinced that the Soviet Union could fight and win a nuclear

---

[11] The United States argued that the invasion was lawful. A review of the legal matters found the rationale to be unconvincing but concluded that, referring to the United States, "ultimately, the United Nations cannot prevent such powers from pursuing what they perceive as in their fundamental national security" (Murphy, 2004, p. 178).

war).[12] Figure 2.1 dramatizes the contrasting errors.[13] Yet another form is when we misunderstand the adversary's beliefs about our own intentions or resolve.[14]

2. Miscalculations, such as the aggressor underestimating the potential negative consequences and the defender overestimating the quality of deterrence. Miscalculations also include errors in military assessments, such as likely outcomes of battles and wars. The errors could be due to omitting important factors, mis-estimating their values, or misunderstanding how to estimate consequences (model error). Miscalculations need not be blunders.

3. The tyranny of the best estimate.[15] All too often, decisionmakers embrace the best estimate uncritically and proceed without

---

[12] Prominent figures argued that the Soviets were insensitive to human losses because they had lost 20,000,000 people in World War II (Wohlstetter, 1959, p. 222) but had survived and won the war. Others argued that, to the contrary, because of the disastrous losses, the Soviets were especially aware of the need to avoid general war (Freedman and Michaels, 2019, pp. 175–177). The Soviet military was indeed far less sensitive to casualties than were NATO nations. The Soviet military also prepared quite seriously for nuclear war, but Soviet leaders were different. Khrushchev and other Soviet leaders came to believe that nuclear war was "unthinkable, if not impossible" (Craig and Radchenko, 2018).

[13] Some of the second class of errors to avoid were discussed in the Cold War (Garthoff, 1978) and updated more recently (Mandel, 2009).

[14] In the Cold War, Soviet military officers were convinced that war, if it came, would be when NATO invaded the Soviet Union or the United States conducted a nuclear strike. More recently, North Korea, Iran, and Iraq have had good reasons to fear U.S. aggression, especially after being identified in 2002 by the Bush administration as part of the "axis of evil"—and even more so if they read a subsequent book by some people close to the administration (Frum and Perle, 2004).

[15] The phrase "tyranny of the best estimate" was introduced in Davis and Arquilla, 1991a. It was discussed later by Charles Duelfer, who led the post-conflict study of mass-destruction weapons in Iraq (Duelfer and Dyson, 2011, p. 99). The authors ask the following:

> How differently would the U.S.-Iraq dyad have evolved if policymakers on both sides possessed a perfect understanding of each other? This is of course the realm of counterfactual speculation, but it does seem reasonable to ask, with Saddam deposed and dead, and in light of a U.S. involvement in Iraq that has been incredibly costly in both lives and treasure, whether either side would

Figure 2.1
Opposed Errors to Avoid: Mirror-Imaging and Stereotyping

Mirror imaging	Stereotyping
He would never do that: It would be wrong and illogical. He would never imagine that because it is unrealistic and our intentions are good.	He is inherently aggressive and militant. He can be deterred only by the certainty of military defeat. He doesn't even understand "deterrence" in the way we do. And he can't be trusted with any agreement.

preparing seriously for other possibilities. It is one thing to misunderstand the adversary and make some miscalculations; it can be even worse to fully embrace those best estimates and

---

have chosen the same courses of action had it possessed a more accurate understanding of the other. (p. 96)

Robert Jervis also mined the history of the ill-fated 2003 invasion (Jervis, 2006a). One of the points he highlighted was "too much certainty" as he lamented failure to look at alternatives seriously (pp. 14–15). He also mentions excessive consensus (p. 21). Jervis emerged from his study pessimistic about whether matters would improve much, despite the many opportunities.

forgo hedging. Anticipating other possibilities is especially important because expert best estimates are often wrong.[16]

Lessons from Cognitive Psychology

Classes of Decisionmaking Error

One of the most fundamental lessons from both history and psychology is that human decisionmaking is only sometimes well described by a rational-actor model. More generally, we should assume limited rationality, even when dealing with heads of state surrounded by advisers and informed by intelligence sources.[17]

Much has been learned in the past 60 years about the strengths and foibles of human decisionmaking—demonstrated not only by researchers testing college students but also by the historical record of high-level political, economic, and business decisions. We touched on some of this in the section about historical failures of deterrence. We seek to apply lessons learned from the research to assess the quality of deterrence and ways to improve it. We see deterrence in terms of favorably influencing the adversary's judgments and decisions. In doing so, we consider perceptions and assumptions that may or may not be accurate, cognitive biases, and the decisionmaking styles of adversary leadership—all these are related and overlapping. For example, the idiosyncratic features of a particular leader may affect all of these.[18]

---

[16] As a related matter, experts—although excellent at identifying relevant factors—are notoriously bad at forecasting (Tetlock, 2017). Significant research has sought to improve forecasting but doing so is challenging and tends to emphasize aggregating results from appropriately diverse estimators (Tetlock and Gardner, 2015). The best forecasters (so-called foxes, rather than hedgehogs) tend to have broad knowledge and open minds, rather than specialized in-depth knowledge.

[17] See the pioneering work of Robert Jervis (Jervis, 1976; Jervis, 2017) and our discussion in Chapter Two of National Research Council of the National Academies, 2014.

[18] The interplay of these factors can be seen in other domains, such as counterterrorism. This domain includes interpretation of U.S. failure to deter Osama bin Laden, who was contemptuous of the United States for its lack of courage and willingness to take casualties (Schachter, 2002, p. 84).

As a separate matter, when we construct U.S. influence strategies, we want to mitigate the potential for errors even though this report focuses on influencing Red. After all, U.S. decisionmakers are also subject to cognitive biases.

Cognitive Biases

The psychological literature discusses scores of distinguishable cognitive biases. In Table 2.1, we present one way to organize them in a taxonomy (Davis, Kulick, and Egner, 2005).[19] The authors noted that the often-maligned cognitive biases are often very useful in practice. Indeed, intuitive reasoning is often more feasible and even superior to rational-analytic thinking. For crisis decisionmaking, however, it is important for decision aids to include so-called rational-analytic thinking prominently to avoid serious errors.

A recent article by Kahneman, Lovallo, and Sibony (2019) stress a subset of biases when advising strategic planners.[20] As discussed succinctly in the article, imposing some structure on decisionmaking can mitigate cognitive biases. The authors refer to their approach as the Mediating Assessments Protocol (MAP). Core elements include (1) defining the assessments in advance (i.e., identifying key factors), (2) using fact-based independently made assessments, and (3) deferring final evaluation until after the mediating assessments are complete, presented, and discussed.

---

[19] This was written at the time of a great scientific battle between the heuristics-and-biases school led by Daniel Kahneman and a "naturalistic" school associated with Gerd Gigerenzer (Gigerenzer and Selten, 2002) and Gary Klein (Klein, 1999; Klein, 2001). Today, the issues are often discussed in terms of fast versus slow reasoning, both of which have their place (Kahneman, 2011). The 2005 study, however, made an additional point, which is that rational-analytic and naturalistic-intuitive styles are useful competitors even when deliberation and slow thinking is possible. Attempts at rational-analytic reasoning often lack creativity. Naturalistic-intuitive reasoning may be more creative and less bound to conventional wisdom and analysis. It can, however, be spectacularly wrong-headed.

[20] Kahneman, Lovallo, and Sibony, 2019, highlights what the article refers to as the availability bias, confirmation bias, excessive-coherence bias, mental-model problems, noise, and the representativeness bias.

Table 2.1
A Partial Taxonomy of Cognitive Biases

	Bias	Description
Memory	Availability	Recent or emotional events are more effectively available or retrievable by memory
	Imaginability	Event seems probable because it is easily imagined
	Representativeness	Event seems more probable if representative of its class
	Testimony	Recalled details may be logical, coherent, and wrong
Naive statistics	Base rate and chance	Normal occurrence rates may be ignored when seeing what appear to be unusual events
	Sample size	Sample size is often ignored when inferring strength of evidence
	Frequencies and probabilities	Equivalent data are perceived differently when expressed in frequencies or probabilities
Adjustment	Anchoring	Assessments are made in relative, rather than absolute, terms, even if baseline is arbitrary
	Conservatism	New information is accepted reluctantly or ignored
	Regression	Events may be overweighted, ignoring likely regression to mean
Presentation	Framing	Events framed as gains or losses are seen differently
	False analogy	Current problem may be seen as like a familiar one, when it is not
	Attribution	Information may be unreasonably rejected or accepted if source is disliked or liked, respectively
	Order	First or last items tend to be overweighted
	Scale	Perceived variability of data depends on scale
Choice	Habit	Option may be chosen for its familiarity
	Attenuation	Decisionmaking may be simplified by discounting uncertainty
	Inconsistency	Judgments may be inconsistent for identical cases

Table 2.1—Continued

	Bias	Description
Confidence	Completeness	Apparently complete data may stop search
	Confirmation[a]	Only confirmatory evidence may be sought and disconfirmatory evidence may be rejected; inappropriate dissonance reduction may occur
	Illusion of control	Sense of control may be unduly enhanced by good outcomes obtained for wrong reasons

SOURCE: Davis, Kulick, and Egner, 2005, Table 2.1, p. 12.

[a] This is closely related to motivated bias, although distinctions are sometimes drawn with confirmation bias, which causes us to see what we want to see. Motivated bias, on the other hand, inclines us to be especially critical of evidence disputing what we want to believe or already believe. Interestingly, motivated bias has been studied at the neural level in functional magnetic resonance imaging (fMRI) studies (Weston, Blagov, Harenski, Kilts, and Hamann, 2006). It has also been studied extensively in international relations. Jervis (2006b) discusses its function of reducing cognitive dissonance. Others emphasize the related function of achieving closure (Bar-Joseph and Kruglanski, 2003).

Prospect Theory

Neither Table 2.1 nor the Mediating Assessments Protocol discussion highlights the most celebrated fruit of Kahneman's research, so-called prospect theory (Kahneman and Tversky, 1979), although it is covered in Table 2.1 as part of presentation > framing.

Prospect theory refers to the empirically established tendency of individuals and groups, when they are in the domain of losses, to make decisions that are bad bets. This might correspond to someone placing a bet in a casino in hopes that he will win and thereby avoid losing the recently purchased sports car that is about to be repossessed. Even if recognizing that he will most likely be even worse off betting than not (he'll probably lose the bet and his car), he takes the irrational risk in the desperate hope that the bet will pay off. In contrast, people are less willing to take risks when the baseline circumstances are favorable (the domain of gains). As a related matter, humans dislike losing something they possess even more than they value gaining the same thing if they do not yet have it (the endowment effect). Unfortunately, for attempts to predict action, the adversary may see himself in the domain of losses even when we believe the adversary is in a good position and should be risk-averse. This may occur if we fail to appreciate grievances or fail to appreciate that, for some individuals, missing the chance for historic glory would be a big loss.[21]

Prevalence

How often do the various problems arise? In Table 2.2, we show our attempt to relate the various errors to the historical cases mentioned earlier. A + sign in a cell indicates that the particular class of error occurred. Many subjective and arguable judgments are needed when making such an appraisal, but we think that the coding is accurate enough to make the point that the problems have been common in past decisions about war. It seems to us likely that extended debate would cause more of the cells to have + signs, strengthening the point.

---

[21] Someone in the domain of losses is likely to become increasingly and "'unreasonably' risk accepting as they become emotionally more dissatisfied with their current situation and trends" (Davis and Arquilla, 1991b, p. 6). Perhaps the earliest application of prospect theory to political-military crisis was in a pre-invasion study of possible aggression by Saddam Hussein in 1990 (Davis and Arquilla, 1991a). Saddam, in 1990, was very upset with his current situation and what he saw as insults, treachery, and plots against him by Kuwait, the United States, and others. Some described him as having a siege mentality (Stein, 1992, pp. 168, 178).

Table 2.2
Types of Error in the Cases of Blunder

Case	Phenomenon			
	Misunder-standing Opponent	Miscal-culations	Tyranny of Best Estimate	Cognitive Biases (Affecting Previous Columns)
World War I (all powers "fell into" war)		+		+
United States was surprised by Pearl Harbor	+		+	+
Japan went to war with United States	+	+		+
UK, France failed to deter Hitler	+			
United States had a near-miss in Cuban Missile Crisis	+	+		
United States prolonged the failed war in Vietnam	+	+	+	+
Israel was surprised by Egypt in 1973	+		+	+
UK was surprised by Falklands attack; Argentines initially did not expect strong UK response	+			
United States was surprised by Saddam's invasion of Kuwait in 1990	+		+	+
Saddam invaded Kuwait, which led to losing a war		+		
Soviet Union invaded Afghanistan and fell into quagmire	+	+	+	+
U.S. war in Iraq ended up being long and bloody	+	+	+	+
NATO was surprised by Russia's actions in Crimea and Ukraine	+			

CHAPTER THREE

Cognitive Modeling and Thinking-Red Exercises

This chapter describes our approach generically. The approach should be useful in analysis, human exercises and wargaming, and combinations. Analytically, we see our method as uncertainty-sensitive cognitive modeling. When thinking more specifically about practical exercises that might be conducted within the U.S. Department of Defense, we refer to *thinking-Red exercises* (exercises that attempt to anticipate Red's potential reasoning). Chapter Four applies this approach more concretely to a China-Taiwan crisis.[1]

As a preamble, we might comment on what is arguably new about uncertainty-sensitive cognitive modeling. After all, numerous methods exist for studying how to deter or otherwise avoid aggression against the United States or its allies. As we define it, *uncertainty-sensitive cognitive modeling* seeks to do all of the following:

- Broaden the framing of the challenge to go beyond traditional deterrence.
- Confront uncertainty about how the adversary is perceiving, reasoning, and deciding—and go well beyond rational-actor thinking or game theory in doing so.
- Encourage defining a robust strategy that can, with adaptations along the way, be effective despite our current uncertainties.

---

[1] The potential of war due to Chinese attack of Taiwan has been of concern for decades. Reasons for concern are being discussed even in recent times (Blanchard, 2020b) and Shih (2020).

- Serve to inform the design of well-structured human exercises and analytical studies.

Challenges and Approach

In developing the approach, we summarized challenges to be addressed (Table 3.1). The items are drawn from the discussion in the earlier chapters.[2] We use the convention that Blue is attempting to influence Red. In a more comprehensive treatment, of course, Red affects Blue, Blue affects Red, and both interact with others in what can be complicated n-party dynamics (multiple players, multiple time periods, multiple branches at each decision point).

Table 3.1 shows that the first challenge is framing the problem well—for example, seeing deterrent threats as merely one instrument of influence. A second is understanding how the adversary may be reasoning. A third is avoiding the tyranny of the best estimate when reasoning because best estimates are often badly wrong. The subsequent challenges are formulating influence strategy that is well hedged in light of uncertainties, implementing the strategy, and likely adapting in the light of additional information.

Figure 3.1 shows our approach schematically. The intent is— referring to the numbered items—to (1) model Red, (2) characterize Red's perceptions and options, (3) anticipate what would constitute the perfect storm in terms of circumstances and the nature and perceptions of Red so as to cause Red to consider major aggression, (4) construct alternative Blue influence strategies that might affect Red's decisionmaking and quell the storm, (5) assess and compare the candidate strategies, and (6) use insights from this to construct a robust strategy, often a hybrid of the initial candidates. As indicated by the icon stacks 🗂, we use alternative models of Red and alternative Blue perspectives in some of the steps to determine whether a given strategy

---

[2] A discussion of tailored deterrence on why and how to use actor-specific knowledge can be found in Appendixes D and E of National Research Council of the National Academies, 2014. These review political science theories (e.g., Walker and Malici, 2011) and methods of psychological profiling (Post, 2008) related to deterrence failures.

Figure 3.1
Overview of the Approach

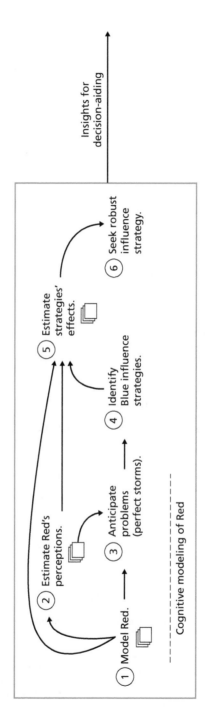

Table 3.1
Itemizing the Challenges

Challenge	Approach
Frame the challenge.	
See challenge as influencing reasoning and choice of human decisionmakers, going well beyond deterrence by threat.	Seek ways to affect the adversary's cognition. Think in terms of influence, not just deterrence by threat. Recognize role of perceptions, emotions, and individual and group characteristics and styles.
Understand how the adversary may be reasoning.	
Understand Red's possible thinking relating to:	Avoid both mirror-imaging and malign stereotyping.
• framing of issues and options • values, interests, fears, and other emotions • perceptions of Blue • other perceptions (military, political, economic, situational) • calculations and miscalculations • reasoning and errors thereof.	Use history, cognitive psychology, leadership profiling, and other theory to recognize where Red errors can occur. Confront and deal with uncertainty in all of the above. Pay attention to Red's writings and behavior, including its attempts at signaling
Anticipate perfect-storm possibilities.	
Avoid tyranny of best estimate when anticipating possibilities.	
Formulate Blue's influence strategy.	
Identify cross-domain influence-action candidates (political, military, economic).	Use history, psychology, and theory to recognize where errors can occur (both Red and Blue are vulnerable to many of the same problems). Strategy should be well hedged and adaptive.
Identify composite strategies.	
Assess possible effects, recognizing uncertainty.	
Compare options, avoiding own errors of perceptions, calculation, and reasoning.	
Choose robust strategy.	
Implement and adapt the influence strategy.	Implement, observe, and iterate according to developments (not addressed in this report).

action will have the desired effect. Thus, uncertainty is included at all stages of the analytical process and reflected in ultimate decision aids.

To implement the approach of Figure 3.1 in addressing the challenges in Table 3.1, we needed an array of mechanisms or tactics. For example, looking at Figure 3.1, how are we to model Red, or anticipate problems, or identify possible influence strategies?

The term *uncertainty-sensitive cognitive modeling* refers to the whole set of mechanisms we discuss in the remainder of this section. The underlying philosophy can perhaps best be understood by contrasting it with familiar baseline mechanisms for thinking about the adversary. Many of these draw on game theory and the version of deterrence theory that assumes that deterrence is achieved if the adversary soberly assesses the benefits and costs of an aggressive option and, after a mental or mathematical calculation, concludes that the aggression is not in his interest.[3] Despite the analytically compelling nature of this formulation, it does not adequately explain the many instances in which deterrence has failed.[4] Nor does not it relate well to the way that humans typically discuss and reason, even at the highest levels of government. Human reasoning and discussion often depend heavily on narratives, which are sometimes helpful and sometimes injurious. These narratives depend on myriad perceptions and judgments subject to cognitive biases and other psychological matters described in Chapter Two.

The "uncertainty-sensitive" adjective in uncertainty-sensitive cognitive modeling applies because thinking-Red is notoriously difficult. Thus, our approach emphasizes entertaining alternative models. Furthermore, it recognizes the need to vary the parameter values of those models, which often relate to perceptions rather than reality.

The mechanisms within uncertainty-sensitive cognitive modeling include the following:

---

[3] This paradigm is common in the deterrence literature (George and Smoke, 1974) and related U.S. military strategic doctrine (U.S. Strategic Command, 2006).

[4] Significantly, major foreign-policy errors have sometimes led not to failures of deterrence but to overreacting to perceived threats (Walker and Malici, 2011).

1. Factor trees. We use qualitative models called *factor trees* to characterize the primary factors thought to affect Red decision-making at different levels of detail. For this work, we seek factor trees that convey plausible Red narratives leading to what Blue sees as aggression. Thus, in this report, they help to anticipate perfect storms, in which a variety of considerations cause Red to see the aggressive action in question as desirable or even necessary. Factor-tree methods have been used in a diversity of studies.[5]

2. Alternative models of Red are what their name suggests. Although unusual, the recognition of such explicitly different models has a long history (National Research Council of the National Academies, 2014, pp. 73–75). Knowing that the adversary has been notoriously difficult historically, with many striking misperceptions.[6] Having alternatives can help but only if they are suitably chosen and taken seriously.

3. Limited rationality. We use models of decisionmaking under limited rationality with potential effects of cognitive errors (some driven by emotions, such as desperation), idiosyncratic reasoning styles, different risk-taking behaviors, and erroneous perceptions.[7] All of these problems are exacerbated by having limited or erroneous information and time pressures.

4. Influence actions and influence strategies. We identify discrete political, military, and economic Blue actions that might be taken and construct composite influence strategies as combinations thereof. These strategies will include deterrent

---

[5] Factor trees were introduced in counterterrorism research (Davis and Cragin, 2009; Davis and O'Mahony, 2013), with Davis and O'Mahony having included a primer in an appendix (pp. 49–67). Factor trees have many applications in defense (Davis and O'Mahony, 2017) and in policy analysis generally.

[6] Keren Yarhi-Milo has striking case histories on this matter, which highlight the differences between how intelligence organizations see adversaries in very different frames than do top policymakers (Yarhi-Milo, 2014).

[7] Such models ware used in anticipating possible behavior of Saddam Hussein's potential behavior before and after his 1990 invasion of Kuwait (Davis and Arquilla, 1991b; Davis and Arquilla, 1991a).

threats but might also include inducements, reassurances, and dissuasive elements.

5. Estimating influence with models of Red. In assessing alternative Blue influence strategies, we estimate their effects using the different models of Red and different Blue perspectives about such effects. Major disagreements commonly exist about how effective a strategy will be, even when analysts are using the same sketchily defined model of Red. We represent the consequences of such disagreements with alternative perspectives that attempt to bound the issue.

6. Seeking robust strategy. We conduct exploratory analysis across the many sources of uncertainty (e.g., different models of Red) to seek influence strategies that are robust—that is, those that might be expected to be effective across the uncertainties.[8] This depends on such strategies being flexible, adaptive, and resistant to shock. Such strategies often allow for iteration and modification, rather than ruling out actions that may prove necessary or persistently applying actions that prove counterproductive. We measure robustness heuristically in this study, but this can be done more systematically using measures of regret and methods documented elsewhere (Groves, Molina-Perez, Bloom, and Fischbach, 2019).

The first three are used in steps 1–3 of Figure 3.1 (and later), and the second three are used in steps 4–6.

Table 3.2 summarizes how the mechanisms relate to the challenges identified in Table 3.1 to indicate that we have sought to be systematic in addressing the challenges. Rather than describe the mechanisms

---

[8] *Exploratory analysis* is multidimensional uncertainty analysis. Whereas the more familiar sensitivity analysis usually varies assumptions one at a time around a best-estimate point, exploratory analysis varies all relevant assumptions simultaneously, taking into consideration their full range of possible values. Thus, exploratory analysis explores a vast case space (also called *assumption space* or *scenario space*). First introduced in defense planning (Davis, 1994), exploratory analysis is a close cousin to exploratory modeling, referred to in the literature on Robust Decision Making (Lempert, 2019). It is also related to what has been called *morphological analysis* (Ritchey, 2011).

in detail here, it is more useful to illustrate them concretely. The remainder of this chapter does so in the following order: (1) factor trees to characterize the factors most important in the adversary's thinking, (2) models of Red, (3) a decision model for representing Red's limited rationality and consequences of cognitive errors (also applicable to Blue), (4) constructing Blue influence strategies using models of Red as a guide, and (5) using uncertainty analysis to fashion robust influence strategies.

Factor Trees

As previously mentioned, factor trees are qualitative models that attempt to dramatize the factors important to decisionmakers. In this report, factor trees should capture the narrative that might cause a nation's leaders to proceed with aggression. Factor trees serve as a format for structuring a more cognitively based approach to assessing Red decisionmaking.

Neutral Factor Trees

Figure 3.2 is our first example of a factor tree—one that is neutral. It presents visually the range of causal factors (not statistical correlations) that might influence a nation's decision about some strategic action. The convention in such trees is that if factor A points at factor B, then more of A will tend to cause more of B—unless a negative sign is attached, in which case more of A will tend to cause less of B. If arrows (arcs) pointing to a node are connected by *Or*, then the node may be dictated by any one of the contributing nodes or a combination. In other factor trees, *And* may appear, which would indicate that, at least to a first approximation, all of the subordinate factors are needed for the target node to be significant.

    Figure 3.2 is neutral in that it just displays considerations without any hint about which might be more important. Factor trees for specific cases can indicate relative emphasis by varying the thickness of arrows.

Table 3.2
Analytical Mechanisms for Confronting Challenges

Challenges	Mechanisms					
	Factor Trees	Alternative Models of Red	Models of Limited Rationality	Influence Actions and Strategies	Estimating Effects with Models of Red	Assessing Robustness of Strategies
See adversary as human.	•					
Go beyond deterrence by threat.	•			•		
Understand Red's thinking (with humility) with regard to	•	•	•			
framing	•	•	•			
interests, fears, values	•	•	•			
perceptions of Blue		•				
other perceptions	•	•	•			
calculations	•	•	•			
reasoning style.	•	•	•			
Anticipate perfect storms.	•		•			
See influence as a multidomain strategy.				•		
Assess potential effects on Red's thinking.	•	•	•	•	•	
Compare influence options, avoiding own errors of limited rationality.					•	•
Choose robust strategy.						•

Figure 3.2
A Neutral Factor Tree from Political Science in the Great-Power Competition

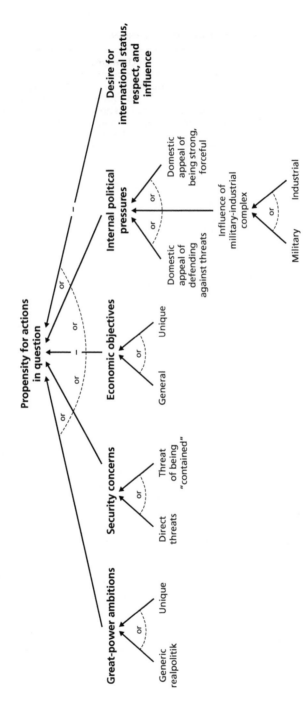

Factor Trees to Convey Narratives and Anticipate Perfect Storms

In this report, we are concerned with influence strategies to avert acts of aggression against U.S. interests. For that purpose, it is useful to employ factor trees that tell a story—that is, that convey the narrative of a potential aggressor to whom an aggressive action may be attractive. A better factor-tree depiction for that purpose is shown in Figure 3.3. Upon inspection, it tells a story that is understandable in a left-to-right reading. The country has motivation for the action. As discussed later, the country may have hope of gaining power, influence, or wealth. However, it may instead have some form of motivation by necessity, such as the threat of losing something important. But would acting on this motivation be legitimate? Red may or may not be concerned about internal support, international support or acquiescence, or various moral concerns.[9] Moving right across the figure, if the motivation is strong and the action is legitimate, are the prospects good, and will the action succeed? Even better, will it be easy to accomplish and relatively painless? By this point in the process of reading left to right, we see considerable tendency toward action.[10]

To be sure, there would be complicated consequences. The rightmost main factor calls for summing up other risks, benefits, and costs. This factor tree allows for a more comprehensive net assessment in which there is a consideration of upside possibilities, downside risks,

---

[9] Readers may be skeptical about decisionmakers having moral concerns when making foreign policy decisions, but in fact they often do. Joseph Nye has recently reviewed how moral considerations have affected U.S. presidents since the 1940s onward. He considers issues of intentions, means, and consequences (Nye, 2020). We are not aware of a similar review relating to foreign leaders. Some authors have discussed related matters in terms of "appropriateness" (March and Olsen, 2011). For example, there seems to be a taboo against the use of nuclear weapons (Tannenwald, 2007).

[10] This cognitive model is intentionally structured differently from, for instance, an enumeration of pros and cons or a cost-benefit calculation. Legitimacy, for example, is an important factor in the reasoning of some national leaders but not others. A cost-benefit calculation, on the other hand, might be reflected only indirectly if at all (e.g., when doubtful legitimacy would imply paying some political cost). Prospects would appear implicitly as part of probabilistic calculations of different possible outcomes. Our claim is that actual reasoning may be more heuristic, as illustrated in the factor tree in Figure 3.3 (Davis and Cragin, 2009). Leaders may also attempt a more deliberate style of reasoning that is closer to decision analysis.

Figure 3.3
Generic Factor-Tree Considering an Aggressive Action

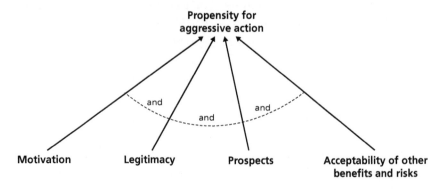

various cases, and so on (right side). But if the leadership's desire to take action is strong enough, shortcuts will be taken and optimistic assumptions will be made that will diminish the impact of the last factor. Also, this particular example assumes that legitimacy plays a lesser role than the other factors.

What we are painting in this factor tree, then, is the kind of narrative that might go with a perfect storm: circumstances in which a potential aggressor has strong motivations (e.g., passions for gain or desperation about not losing something important); a sense of legitimacy; importantly, a sense that the operations probably will go quickly and decisively with relatively little pain (or, at least, pain well worth tolerating given the prize);[11] and, lastly, a sense that other considerations do not change the picture much.

---

[11] John Mearsheimer highlighted this point decades ago when arguing that NATO's conventional deterrence of the Warsaw Pact was most likely to fail only if a relatively quick invasion appeared possible to the Soviet planners (Mearsheimer, 1983).

Models of Red

An important element of our approach—avoiding uncritical embrace of the best estimate—is having alternative models of Red. Such models can be simple or complex, small or large. They should reflect insights from the literature of country specialists and be informed by both rational-actor and limited-rationality theories, by lessons from history, and, significantly, by the analytic purpose of covering the range of not-implausible behaviors. At the same time, the models need to be relatively simple.

For this experimental work, we use the dimensions summarized in Table 3.3. The first is the strength of the motivation that Red may have for what Blue regards as aggression.[12] This strength depends on Red's national interest, ideological and political factors, and (too often neglected) psychological traits of relevant individuals and groups (National Research Council of the National Academies, 2014, p. 121). The second is what we call *decisionmaking style* (e.g., intuitive

Table 3.3
Illustrative Dimensions for Defining Red

Dimensions	Values
Motivation: strength of interests	Low, moderate, or high (as with protecting vital interests)
Decisionmaking style	Intuitive and perhaps impetuous, rational-analytic
Risk-taking propensity	Low, moderate (will take good bets), high (as when desperate or when driven by upside and oblivious to risks)
Realism	Too optimistic, accurate, too pessimistic
Resolve	Low, moderate, high
Red's view of Blue's intent (Red's Blues)	Any of itemized Red's Blues (see Table 3.5)

---

[12] This discussion treats the nation like an individual actor, despite complex multiparty interactions. The approximation might apply if decisionmaking is dominated by a dictator but might also apply in a non-unitary government if the nation's decisionmaking behavior is akin to being made by a unitary actor. Cautious, risk-avoiding behavior, for example, could reflect disagreement and paralysis rather than a dominant actor being cautious.

versus rational-analytic). The third is risk-taking propensity. By *risk taking*, we mean taking a bad bet relative to one's own values. A nation that starts a war from which it will gain modestly with 55 percent probability but will lose greatly with 45 percent probability is taking a bad bet. Because the bets at issue are very consequential (e.g., wars), a nation with low risk-taking propensity would avoid actions that might lead to very negative consequences even if a favorable outcome would be somewhat more probable. In contrast, a nation with medium risk-taking propensity would take those actions. A nation with high risk-taking propensity would take actions if it seemed reasonably possible that it would achieve great gains, despite the downside risks associated with failure. Such a nation may have a leader who is impetuous and does not pay much attention to risks. This can easily happen when the leader is in the domain of losses or, interestingly, when a leader sees himself or herself as a great conqueror or a great leader of historical significance. Such a leader does not think of the expected gain over a thousand variants of the same scenario; this leader only lives once and does not wish to lose a once-in-a-lifetime opportunity for glory. The same can be said of many entrepreneurs. When characterizing Red's risk-taking propensity, we distinguish between normal circumstances and the possible circumstances of crisis in which leadership might be exuberant about potential gains because they are desperate about an imminent perceived threat (external or internal) or otherwise emotional and perhaps risk-taking.

A fourth dimension for characterizing Red is Red's realism in perceiving the world and making assessments, to include estimating the chances of winning a conflict. Is Red accurate in such matters, whether being optimistic or pessimistic? Given an ambiguous military situation, for example, would Red assess a potential conflict as dicey (accurate), as one that it would most likely win (optimistic), or as one that it would most likely lose (pessimistic)? Red's decision at any point in crisis or conflict would also depend on its *resolve*—the firmness of its determination.

The last feature is how Red views Blue's intent. In particular, does it seem to Red that Blue considers Red to be a mere competitor or as an enemy to be vanquished? What does Red think about Blue's resolve?

Does Red see Blue as willing to fight, or is Red's view that "Blue doesn't have the stomach for this and will back off rather than fight"?[13]

This list of dimensions is neither unique nor equally good for all problems. It is what we arrived at after considerable discussion for the current study.[14]

Table 3.4 shows an example of how alternative Red's Blues might be characterized succinctly. The labels RB1 and RB2 are generic placeholders, but, in this example, participants decide the characteristics.

It is important to recognize here that even a rational-analytic Red may be comfortable taking risks that Blue sees as reckless but that Red reasonably sees as good bets. In Table 3.4, we use the term *risk-taking* to mean willingness to take risks that are not good bets.

An important element of our understanding of Red is Red's model of Blue. For this, we use the same dimensions as in Table 3.3 but substitute Blue's Red for Red's Blue. Table 3.5 shows some illustrative Red's Blues. Participants in an exercise could add or subtract columns if necessary, but we see that the dimensions, which provide significant diversity, will often be useful.

Table 3.4
Illustrative Alternative Red's Blues

Model	Label	Characteristics
RB1	Sensible	Distinguishes between interests and vital interests; is rational-analytic in style, with a tilt toward caution; even in crisis would be deliberate and rational
RB2	Risk-taking	Tends to see interests as vital; attempts to be rational but can be opportunity-seeking and impetuous, overemphasizing upside potential and underestimating or ignoring downside risks

[13] One example of this was Saddam Hussein's belief that the United States had no taste for ground warfare and thus lacked staying power (Stein, 1992, p. 175; Schachter, 2002).

[14] A much longer set of adversary characteristics was developed years ago when researchers sought to represent alternative models of the Soviet Union during the Cold War (Kahan, Schwabe, and Davis, 1985).

Table 3.5
Illustrative Red's Blues

Red's Blues	Blue's Motivation	Blue's Decision-making Style	Blue's Risk-Taking Propensity	Blue's Resolve	Blue's Image of Red
RB1: Rational risk-averse competitor	Moderate	Rational-analytic	Low	High	Competitor
RB2: Driven intuitive threat-perceiving	High	Intuitive	High when desperate or otherwise emotional	High	Enemy
RB3: Rational but with low resolve	Moderate	Rational-analytic	Low	Low	Enemy

Representing Limited Rationality

We have developed two methods for representing Red's attempt to make sensible decisions. The first is largely heuristic and tied to Red's factor tree. The method depicts how Red combines the top-level factors of its factor tree to assess the attractiveness of an option, building a narrative in the process. The second method is more deliberate and more like rational-analytic thinking. However, as we will illustrate, it can represent common errors of reasoning related to cognitive biases. Figure 3.4 shows both schematically for a generic application. The rightmost branch (in the dashed box) refers to the more-deliberate mode of decisionmaking. The branches to the left correspond to the factor-tree approach.

Heuristic Factor-Tree Method

Much decisionmaking consists of leaders musing about considerations and then reaching a conclusion directly. Many experts, when asked to make a judgment, will consider the relevant factors, roll them over in their heads, and make their judgment. The first mechanism for aiding such decisions is to be systematic in reminding the decisionmakers about the factors they should be considering. In applying our approach in an exercise, for example, the Blue team attempting to estimate Red's

Figure 3.4
Red May Reason Heuristically or Deliberately

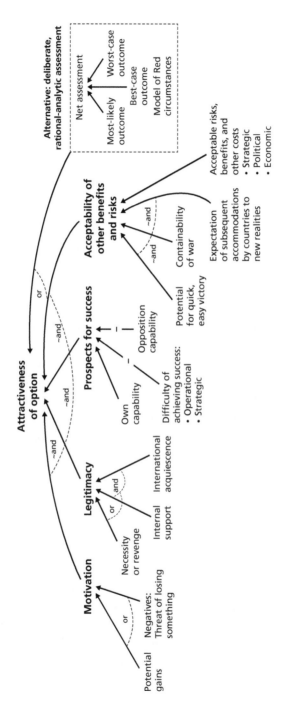

assessment of options (for each model of Red) can be reminded of the factor tree and encouraged to think about each of its factors.

A variant mechanism is to employ numerical scores and mathematics. With this, participants estimate scores for each of the top-level factors in the tree using a 0 to 10 scale. A 10 means that the factor is very favorable to the option in question, whereas a 0 means that it is very unfavorable (e.g., if prospects for victory are very poor). The factors are then combined mathematically to generate an assessment of the option. We suggest using the method of thresholded linear weighted sums, which incorporates the common heuristic of discarding (filtering out) options that fail to reach a threshold of unacceptability for one or more of the factors considered critical to the option. We describe this in Appendix A; additional details are available in other research (Davis and O'Mahony, 2013).

Deliberate Method Highlighting Outcome Uncertainty

A more deliberate method of assessing an option attempts to account for probabilities. For a given option, the method considers the relative probability of all possible outcomes and then estimates the expected value (i.e., the mean outcome) by summing or integrating across possibilities and their probability densities. This is the familiar method of Decision Analysis 101. The expected utility U_i of the ith option is the sum over the possible outcomes of the probability P_j of the jth outcome multiplied by the utility of that outcome O_{ij}.

$$U_i = \sum_{j=1} P_j O_{ij}$$.

The allegedly best option, then, is the one with largest utility.

Realistically, people are only seldom able to make such calculations. They lack the information needed and do not know the probability distributions. Additionally, estimating outcomes may require complicated models with many unknown elements, including the very structure of the model. This is the problem of bounded rationality discussed by Nobelist Herbert Simon in the 1950s (Simon,

1957; Simon, 1978). We represent this simply. The concept is that Red is attempting to be rational in choosing an option suitable for reaching its objectives.

Red considers a variety of options and evaluates them by considering a number of factors (as just shown in the factor trees). Furthermore, by recognizing uncertainties in the factor values, the way the world works, and events that may occur, Red evaluates the options for what it perceives to be plausible most-likely assumptions, best-case assumptions, and worst-case assumptions. Red then decides. This structure is rational, but the quality of the decision depends on Red's recognizing all the important factors, correctly perceiving factor values; addressing uncertainty by characterizing outcomes for most-likely, best, and worst cases; and accounting appropriately for this range of outcome uncertainty.

Because it is difficult for real-world decisionmakers to do better, we approximate the calculation rather crudely as a linear weighted sum on what seems most likely, the best that can be reasonably hoped for, and the worst that should reasonably be considered.

$$U_i = \sum_{j=1} P_j O_{ij} \approx W_{ML} O_M + W_{BC} O_{BC} + W_{WC} O_{WC} .$$

This equation represents an intuitive approximation akin to what real people do, but it is mathematically defensible only under special conditions.

Some personalities will heavily weight the most likely case (ML) without much thought to the best case (BC) and worst case (WC). Others will seek to balance the cases so that, in effect, they are attempting to maximize the subjective expected outcome (as in classic decision theory). Still other individuals will focus on the upside (i.e., the best-case outcome) or on the downside risk (the worst-case outcome).

With this background, when we refer to someone's propensity for risk-taking, we have in mind the degree to which the decisionmaker gives too little weight to the downside risk. A conservative style gives high weight to the downside risk, possibly even forgoing an option despite the upside potential, because it would be a "good bet" if we were talking about a recreational gamble without too much at stake. Most people are conservative in real life when it comes to certain important matters because the disutility of losing is so big and they only live once. When considering crisis decisions by nations, they believe that leaders will and should be chary of actions that might lead to nuclear war. A given leader, however, might believe that, even if nuclear war occurred, it would remain very limited and that he or she would "win" the resulting conflict.

Another major source of error is that decisionmakers misestimate the likelihoods of the most-likely, best, and worst cases. Real people use heuristics in such reasoning. They may err in several different ways. Highly conservative individuals may fail to appreciate how good the upside potential of an option is; forward-leaning individuals may tend to dismiss as implausible downside possibilities that would, in reality, sometimes occur and be exceptionally dire. Decisionmakers often put too much weight on the alleged best-estimate outcome, not recognizing that the "best estimate" scenario is actually improbable and its characterization may be a bizarre compromise mixture of optimism, pessimism, and speculation.

Table 3.6 illustrates how the method works in practice using purely notional strategies, Red models, weights, and assessed outcomes. For each alternative model of Red (A and B), we estimate subjectively how Red might rate each of three options for its most-likely, best-case, and worst-case outcomes.[15] Each outcome is scored on a 0 to 10 scale from catastrophically bad to gloriously good. We then assess the options using linear weighted sums with the weights suitable to that model.[16] The method, then, is an analytically rigorous version of what might transpire in a spirited discussion of a leader with his or her senior advisers. One adviser might insist that pursuing an option

---

[15] These outcome estimates could be based on wargaming, simulations, or studies, but expert participants in an exercise will often be able to make the estimates subjectively with little trouble.

[16] In practice, we use thresholded linear weighted sums as shown in Appendix A because some decisionmaking is nonlinear, as when someone rules out options that fail by some criterion.

Table 3.6
Judgment and Decisionmaking with Limited Rationality

Option	Outcome Model A (e.g., rational, unflappable)			Outcome Model B (e.g., risk-taking, impetuous under emotion)			Model A's Net Outcome Assessment		Model B's Net Outcome Assessment	
							Weights (most-likely, best-case, worst-case)			
							Normal	Emotional	Normal	Emotional
	Most Likely	Best Case	Worst Case	Most Likely	Best Case	Worst Case	2:1:1	2:1:1	2:3:1	2:4:0
1. Diplomatic settlement with no gain or loss	3	5	1	3	8	2	3	3	5.3	5.1
2. Some military action	4	6	2	3	8	2	4	4	5.3	6.7
3. Major military action (e.g., invasion)	3	7	0	7	9	2	3.3	3.3	7.2	8.7

NOTES: Participants in an exercise estimate values of the clear outcome cells on a 0 to 10 scale (extremely bad to extremely good) for the circumstances of the specified scenario (these are shown as red entries). They also specify the weights that each model gives to the most-likely, best, and worst cases. The values in the gray cells are then calculated.

would lead to a particular excellent result, albeit with some risk of falling a bit short. Another might argue that, to the contrary, the risk is immense—far too large to be tolerated. The user inputs the items in red (the scores for each option's outcomes and the weights to be placed on the different cases). The shaded area shows the result calculated from the inputs. Note that models of Red may differ in their estimates of most-likely, best-case, and worst-case outcomes and in the weight that they give those different cases in forming a net assessment. For the purely notional example, Model B (shown as being more risk-taking and impetuous, at least when emotions come into play) rates Option 3 most highly, especially under emotional circumstances, whereas Model A favors either very limited military action or none at all.

If significant disagreements exist about the entries in Table 3.6 for a given model of Red, the results can be shown for separate perspectives (as we illustrate later). Doing so leads to more complicated tables.

Influence Actions and Strategies

Ultimately, our intent is to suggest Blue strategies that might plausibly affect Red's thinking. Strategies should address political, military, and economic instruments. There is a need to distinguish between discrete elements of a strategy and the strategy itself. Also, if the approach is to have an analytical character, we need to have some structure so that strategies can be readily compared.

Candidate Actions for Influence

A brainstorming session, human exercise, or wargame could suggest a number of discrete actions to influence Red. These should be guided by insights gained by looking at the factor trees for Red and alternative models of Red. Table 3.7 shows a small set of generic tactics; however, in an application, a richer table would sometimes be necessary. Most of the tactics are traditional, but we added a row to conspicuously include an option with what Blue sees as positives, because, as noted earlier, successful influence usually requires more than threats. It may require inducements (e.g., release of impounded funds as part of consummating the nuclear agreement with Iran), reassurances, and various auxiliary actions to strengthen the coercive efforts. In the example, we show actions in political, economic, and military categories, but more categories could be included.[17]

---

[17] Various groupings have been used, including pol-mil (actually referring to political, military, and economic), DIME (diplomatic, informational, military, and economic), PMESI (political, military, economic, social, and infrastructure), and PMESII-PT (political, military, economic, social, information, infrastructure, physical environment, and time). Some have lamented category inflation.

Table 3.7
Generic Candidate Actions

Category	Action	Description
Political	Diplomatic statements	Expressions of concern, demarches
	Strong diplomatic action	Firmly expressed threats
	Proposal with incentives or reassurances	Diplomatic offers, perhaps secret
Economic	Economic sanctions	Limitations on trade or use of international financial system
Military	Show of force	Redeployment of U.S. forces to troubled region
	Strong military actions	Raising readiness levels and adjusting force postures for actual combat

Influence Strategies

Holding constant the set of possible tactics shown in Table 3.7, we can now define alternative Blue influence strategies by indicating with 1s the tactics to be included. Table 3.8 illustrates this by identifying four possible Blue strategies. Such a set of alternative strategies can be readily developed in a group exercise.

Evaluating Influence Strategies

To assert that an action would usefully affect one of the factors in Red's decision does not make it true. Usually, disagreements exist about probable effects. For example, some senior advisers to a Blue leader might believe that actions would improve deterrence. Others might argue that some of the actions in question have already been assumed and discounted by the adversary, while other actions would be seen as aggressive, threatening, and indicative of a belief that war is inevitable. The effects might then be to raise the adversary's motivation for strong action and undercut arguments within the adversary government about the need for restraint. Because such disagreements are both common and important, we reflect on them with different

Table 3.8
Generic Influence Strategies

	Political			Economic	Military	
Strategy	Diplomatic Statements	Strong Diplomatic Action	Incentives, Reassurances, or Dissuasive Actions	Economic Sanctions	Show of Force	Strong Military Actions
1	1				1	
2	1				1	
3	1	1		1	1	1
4	1		1	1	1	1

NOTE: Red entries indicate inputs by participants.

perspectives (different Blue assessments of the likely consequences of discrete actions or multiaction strategies).

Direct Estimates

One way to estimate the effectiveness of alternatives in an exercise is to have participants discuss the effects they anticipate directly and rather ad hoc, focusing on a particularly aggressive Red option.

Group Estimate

Using the skeleton of Table 3.9 as a template for group discussion, participants could use + or ++ to indicate that a Blue strategy would increase Red's aggressiveness, - or -- to indicate that it would decrease it (the desired effect), or leave the cell blank if they expect no effect. They might be viewing the factor tree as they think about the issue. They would distinguish between Models A and B of Red in doing so. Lively discussion and disagreement might ensue, so note-taking would be needed. If, as in the example, strong disagreements existed among participants, a revised table would record those as alternative perspectives, as in Table 3.10.

Aggregating from Individual Estimates

In an alternative process, Blue participants could fill out Table 3.9 individually, and the exercise control team could process results. If

Table 3.9
Direct Estimates of Effects on Red's Enthusiasm for Aggression

Blue Strategy	Model of Red		Notes on Discussion
	A (Sensible)	B (Risk-Taking)	
1. Minimum			
2. Show force			
3. Strong	-	+	Some see stronger effect on A and negative effect on B
4. Strong + positives	--		

the control team discovered a sharp difference in results that merited being highlighted, they might generate Table 3.10 showing the two perspectives.

For the example, one portion of the Blue team (with Perspective 1) has concluded that a strong Blue option would actually enhance Red's enthusiasm for war if Model B of Red is accurate (remember that - is good), but that the option should be effective for Model A. The other portion of the team (with Perspective 2) estimates that the strategy would be effective in either case. Both perspectives think that the

Table 3.10
Revised Estimates Recognizing Two Blue Perspectives

Perspective	Perspective 1		Perspective 2	
Model of Red				
Blue Strategy	A (Sensible)	B (Risk-Taking)	A (Sensible)	B (Risk-Taking)
1. Minimum				
2. Show force				
3. Strong	-	+	--	-
4. Strong + positives	--		--	

option that includes positives (e.g., reassurances or incentives) might be effective with Model A but not Model B.[18]

More-Systematic Assessment Using Factor Trees

Parallel to the methods used in the section titled "Heuristic Factor-Tree Method" earlier in this chapter, a more systematic but still heuristic variant of such an exercise would ask participants to assess effects of each strategy on each of the top-level items in the Red factor tree as in Table 3.11.

Provisionally, without the benefit of experimental exercises on which to draw, we believe that the simpler approach of Table 3.9 will often be adequately effective and economical. Discussion of the Blue participants might be improved with a decision-aid graphic reminding them of the factor tree for Red. This might encourage participants to use the concepts in discussion, such as saying

> Well, the "strong" option would get Red's attention, for sure, but I'm not sure what effect it would have. If Red were sensible (Model A), it

Table 3.11
Illustrating Group Direct Assessment of Red's Most Aggressive Option

Blue Strategy	Motivation		Legitimacy		Prospects		Acceptability of Other Benefits and Risks		Net Effect (estimated directly)[a]	
Model of Red	A	B	A	B	A	B	A	B	A	B
1. Minimum										
2. Show force										
3. Strong	-	+			-		-	+		+
4. Strong positives	--		--		-		-			-

[a] Participants do their own mental aggregation. As an alternative, participants may enter numerical effects and discuss the relative weight of factors. We are skeptical about the value of this in an exercise. Red entries indicate inputs by participants.

---

[18] Appendix C describes a simple computational model illustrating Bayesian updating of Red's model of Blue. Such a model could be used to integrate insights from human exercises or, in principle, could be used as a decision aid in exercises or studies.

would see prospects as dropping and acceptability of other benefits and risks as worsening as well. However, for a more hot-headed and risk-taking Red, our actions might be seen as provocative and preparation for war. That would increase his motivation for action, especially since he probably believes that war is inevitable anyway.

As for adding incentives or reassurances, the "sensible" Red (and that is who we may well be dealing with) would jump on that if it would suggest a way out of crisis and, conceivably, resolution of long-standing issues. The more hot-headed, risk-taking Red would probably regard our suggested incentives and reassurances as mere fluff and ignore them. Currently, there is a great deal of distrust undercutting our recent attempts at diplomacy.

Deliberate Estimates Recognizing Outcome Uncertainties

A more structured approach is possible. This approach would be Blue considering each model of Red and attempting to estimate, for each model of Red, each Red strategy, each Blue strategy, whether emotions are running high, and how Red would estimate its most-likely, best-case, and worst-case outcomes. Simple calculations could then generate net assessments. This is feasible for an analyst team working over a period of time, but it is tedious and error-prone because of the many combinations of Red, Red strategy, Blue strategy, Blue perspective, and the presence or absence of special emotions. It is likely not worthwhile for a group exercise, unless greatly simplified to address only one or a few of the combinations. We provide an example in Chapter Four.

Seeking a Robust Strategy

The Blue team should examine how each strategy would fare across not only models of Red, but importantly, different perspectives within the Blue team (or in the Blue national security community). The team should then choose the most robust of the strategies or construct a variant that is as robust as it is feasible. In practice, this could mean including such elements as diplomatic initiatives or military actions that would give Red a way out of a bad situation or, conversely, taking stronger military preparatory actions because Red's intentions may be more malign than currently believed. Squaring these might require some finesse because even a contingent diplomatic offer could be seen by Red as a sign of weakness, and even defensive military actions could be seen by Red as escalatory or indicating intention for war.

It is possible that a seemingly robust strategy will arise naturally in a group exercise, somewhat analogous to how dry-wash sessions at the end of wargames sometimes generate consensus and clear-cut alternative insights.

In our view, finding a robust strategy is best accomplished by an iterative process in which trusted analysts discuss issues with policymakers, examine the alternatives with exploratory analysis,[19] use both their intuition and computational methods to find attractive options, and return to policymakers for discussion. Such iterative work is not without precedent.[20]

---

[19] Systematically examining what-ifs is aided by having a computational model that mirrors the qualitative character of a human exercise. Appendix B describes briefly a first-cut computational model that we developed for this purpose. It illustrates feasibility, somewhat mirroring a more fully developed example from another study (Davis and O'Mahony, 2013).

[20] Related work developing alternative military strategies under uncertainty benefited from iterative work with senior military officers (Davis, Gompert, Johnson, and Long, 2008). Tight iterative cycles of analysis and discussion with policymakers has also been successful in the British Ministry of Defense (private discussions with Robert Solly).

CHAPTER FOUR

Illustration: Hypothetical Crisis with China

Against this background, we simulated an exercise applying the approach—for purely illustrative purposes—to a hypothetical crisis in which China is considering invading Taiwan.[1] Our work was episodic over several months but included interactive team sessions following the approach to develop or review factor trees, discuss alternative models of China, and evaluate strategies. What follows illustrates what human exercises might generate. Much more can be accomplished with a combination of systematic analysis and human exercises. **It should be emphasized that this chapter is based on the scholarly literature and independent notional exercises that we and our RAND colleagues performed. In no way does it represent the thinking of the Office of the Secretary of Defense or other official sources. It does not even purport to represent a consensus of RAND analysts. Our purpose was methodological.**

As a road map to the following description of a simulated exercise, we proceed as follows in the same sequence as we did in Chapter Three. This includes (1) presenting an initiating scenario, (2) developing a factor tree for China in the context of a crisis with Taiwan, (3) developing alternative models of China, (4) estimating how each alternative model would assess its options (first in a heuristic, narrative-style approach guided by the factor tree developed earlier and then in a more deliberate approach attempting to balance risks

---

[1] The U.S. challenge regarding Taiwan is an example of extended deterrence, which is difficult because the deterrent threat is inherently less credible. Recent discussions of the Chinese threat to Taiwan include Blanchard, 2020b, and Shih. 2020.

and benefits), (5) developing alternative U.S. influence strategies based on insights gained, (6) estimating effectiveness of those strategies by estimating how the alternative Chinese models would reassess their options in the light of U.S. actions, both heuristically and deliberately, and, finally, (7) discussing what might constitute a robust U.S. strategy that might be effective across the range of plausible models of China.

Initiating Scenario

For our simulated exercise, we could assume prior familiarity of participants with the concepts of our approach. No overview briefing was necessary. Initially, we just discussed context informally and proceeded. Later, we wrote up an initiating scenario with the assumptions that we were using.

(Time of exercise: May 2024)

We are having today's exercise because a crisis is brewing between the United States and China regarding Taiwan. The Taiwanese government is very nationalistic, and it is strongly supported by its public—in part because of China's efforts to marginalize Taiwan by pressuring nations to drop any diplomatic recognition of Taiwan, pressuring international organizations to avoid giving Taiwan a seat at any table, and pressuring international corporations. The crisis is also because of negative reactions to China as the result of the 2020 COVID-19 pandemic, during which China did not discharge its international duties and instead sought to hide early developments of the epidemic. The possibility exists that Taiwan will announce independence soon. That may precipitate a major crisis and might mean Chinese military action against Taiwan, even invasion. China has built or adapted enough ships to support such an operation. And, of course, China is capable of such actions as blockades and missile attacks. You all have your own sources of information, but even public information indicates that China's leadership is very concerned, potentially to an irrational extent, because, over the years, that leadership has centered so much attention on the

Taiwan problem. It may be that China's leadership fears its very legitimacy will be judged by how matters develop. But we don't know.

Your challenge today is to think about ways for the United States to deter Chinese aggression against Taiwan. Unfortunately, we are stretched militarily because of the ongoing conflicts in the Middle East. Indeed, despite years of announcements about shifting resources to the Pacific, we have not actually done much. We are not seriously maldeployed, but we also are not brimming with ready capability in East Asia. In this exercise, you are to think through possible U.S. strategies using a new approach that encourages you to "think Red," but with alternative models of Red in mind, to consider all instruments of power, although with primary emphasis on what the Department of Defense and its military allies can do and to seek strategies that hedge against our many uncertainties.

Factor Trees for China If Considering Acts of Aggression

Our first effort in the simulated exercise was to build a factor tree. One of us developed and presented an intendedly sensible strawman,[2] after which group discussion led to substantial modifications, often in real time. The team was small but included expertise in strategic planning, political science, Chinese military thinking, and modeling. The factor tree evolved over numerous sessions, each being one to two hours long. Figure 4.1 shows the result for the option of outright invasion of Taiwan (lower-level factors would be different for other options).

In Figure 4.1, the top node is a qualitative variable—the attractiveness of a given option for attacking Taiwan. The first level of factors is, as described in Chapter Three, motivation, legitimacy, prospects for success, and acceptability of other benefits and risks. Each of these top-level factors depends on lower-level factors. For example,

---

[2] We chose this approach because results of unaided group brainstorming are often poorly structured and conceived. A strawman is very helpful as a starting point.

Figure 4.1
Factor Tree for China Contemplating Invasion of Taiwan

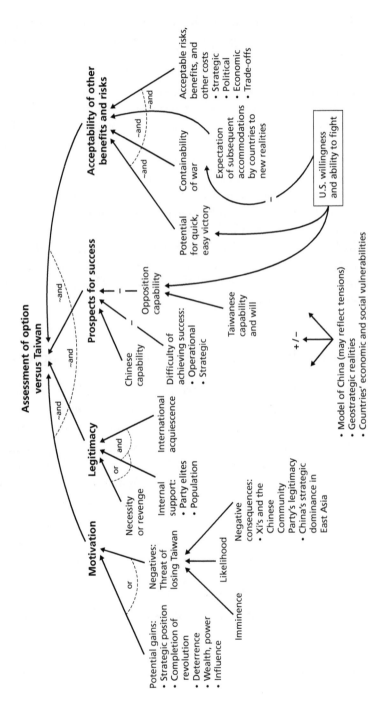

motivation depends on the positives associated with the option, such as the potential for strategic gains, completing the revolution that began with Mao, deterrence (i.e., making it more difficult for the United States to attack China), wealth and power, and influence in the region. Motivation also depends on negatives, such as on the *threat* that China might feel from Taiwan declaring independence. How big the sense of threat might be would presumably depend on how soon Taiwan's declaration might occur, the likelihood of a declaration, and specific negative consequences, notably that the very legitimacy of Chinese President Xi Jinping and the Chinese Communist Party would be called into question internally because the Taiwan issue has been emphasized for decades by the Chinese leadership. For Taiwan to become an independent sovereign would clearly be a big setback (a big *loss*) with respect to the goal of Chinese dominance of East Asia.

Moving rightward, legitimacy might be judged by the objective necessity to prevent Taiwanese independence or a possible requirement for revenge (not, we believe, relevant to the China versus Taiwan case). Or it might be judged simply in terms of whether an aggressive move would be regarded as appropriate internally and tolerable by the international community.

Moving rightward again, prospects for success depends, of course, on whether China could execute the option in question, how difficult it would be to do so, and how much opposition would be encountered. It also depends on whether success of the operations in question would achieve China's objectives. For example, various successful operations involving cyberwar, covert actions, and gray-area actions might increase Taiwanese and U.S. determination rather than undercut it. The lower nodes shown in Figure 4.1 are actually for full-out invasion. Prospects for invasion would depend on Chinese, Taiwanese, and U.S. capabilities, Taiwanese resolve, and U.S. ability and willingness to fight. The latter would likely loom very large in China's thinking.

The last factor, acceptability of other benefits and risks, would again vary with the option considered. For the invasion-of-Taiwan option, a key would be the perceived likelihood of a quick and easy victory and its complement—the risk that the war would be long and costly. It would also depend on whether the war could be contained and not on extending to more general and perhaps all-out war with the United States. Again, this would depend on U.S willingness and ability to fight (e.g., whether the United States is distracted by war elsewhere or whether U.S. domestic support would exist for war). Moving rightward, other considerations would come into play, such as a variety of costs, benefits, and risks not already covered.

At the bottom of Figure 4.1, we are reminded of other considerations that do not fit neatly into any one of the above branches.

Figure 4.1 as a whole includes the general features described in Chapter Three. The figure supports a narrative consistent with a perfect storm, sees motivation in terms of both potential gains and averting painful losses, sees legitimacy as potentially about perceived necessity and support rather than legalisms, highlights prospects for success (perhaps quick and easy), and treats other considerations of benefits and risks. Implicitly, the basic tree assumes discussion using best-estimate values for most of the various factors. Somewhat awkwardly (right side), the figure allows for a fuller assessment of pros and cons. The thrust of the narrative conveyed, however, might be the left-to-right story: We need to do it (invasion), it would be legitimate to do so, prospects for success are quite good, and, although there are many risks and details to consider in principle, they seem tolerable on balance. *To be sure, if China used such a tree today, we do not believe that it would conclude with that narrative, but our intent was to characterize a plausible future perfect storm.*

We regarded this part of our simulated exercise to be successful: Building the factor tree led to far-reaching substantive discussion. Along the way, we sometimes had more layers and nodes, more distinctions, and more subtleties, but retaining those did not pay its way. Significantly, this was merely a prototype experiment. Other exercises would have generated somewhat different trees, and it would be very useful to compare and learn from their differences.

Alternative Models of China

In this section, we describe two approaches to developing alternative models of China. First, we abstract from the scholarly China-watcher literature, to identify some significantly different models of China. We then take a more analytical approach, as described in Chapter Three. We seek a set of models that bounds the possibilities adequately for our study, reflecting not just the China-watcher material but also possibilities suggested from historical experiences, psychological considerations, and strategic logic.

Insights from China-Watcher Literature

In constructing alternative models of Red, we first looked to the strategic literature, particularly those of China watchers. In Table 4.1, we sketch four such distinguishable models. Based on the literature rather than the constructs in Chapter Three, the table characterizes the models in terms of primary motivation for relevant foreign-policy actions, willingness to escalate tensions, desire to avoid major conflict with the United States, concern about maintaining social stability, and the nature of estimates and assessments and related risk-taking propensity.[3]

The **realist's black box** model views China as behaving as would be expected based on the realist theory in political science (e.g., Mearsheimer, 2003). It sees a coolly rational motivation to maximize both power and security. Such a China may escalate the level of crisis if that appears sensible and would be willing to engage in conflict if it believes that doing so is likely to improve its strategic position. The desire to avoid major domestic problems is a constraint but is not necessarily binding if the right opportunities arise.

The **stability-seeking** model views the Chinese Communist Party as concerned about its ability to remain in power and cautious when it comes to any action that could lead to conflict and disrupt regional stability. In particular, a stability-seeking China would want to

---

[3] These can be seen as representing competing strands of thought recognizable within China.

Table 4.1
Alternative Models of China from the China-Watcher Literature

Model	Motivations	Willingness to Raise Tensions	Desire to Avoid Major Conflict	Desire to Avoid Domestic Economic or Social Turmoil	Accuracy of Assessments of Blue's Intentions and Capabilities
Realist's black box[a]	Maximize power and security opportunistically	Moderate	Low if victory is probable, high if victory is improbable	Moderate-low, depending on how it will impact national power	Broadly accurate
Stability seeking[b]	Satisfy party officials; maintain economic growth and social stability (with less concern about national honor)	Moderate-low, depending on possibility of further escalation	Very high, unless failure to engage in major conflict would cause widespread social unrest	Very high	Accurate, possibly too pessimistic
Xi is going to make China great again[c]	Reclaim China's dominant position in East Asia; outside of East Asia, match or surpass the United States in influence; readjust global norms so that China is seen as the model other countries want to follow	High	Moderate or even low if China believes it has a chance of winning or limiting the United States to, at best, a pyrrhic victory	Moderate	Possibly too optimistic; decision to escalate may be independent of assessment of the United States
Collapsing China[d]	Stave off massive social upheaval or the overthrow of the Chinese Communist Party, perhaps by whipping up nationalist fervor or posting some wins	High	Moderate-low	Very high	Possibly too optimistic; decision to escalate could be for domestic reasons, independent of China's model of the United States

NOTE: Abstractions are based on insights from the references listed below, among others; the references we cited sometimes have nuanced discussions and do not necessarily subscribe to one or another of the models.
[a] Mearsheimer, 2003.
[b] Blackwill and Campbell, 2016, pp. 28–29; Doshi, 2019.
[c] Schrader, 2020; Zhang, 2013, pp. 106–110; Zhang, 2014, pp. 5–19.
[d] Beckley, 2019; Copeland, 2019.

avoid anything that would reduce the rate of economic growth, which could endanger party legitimacy. Although China might take some aggressive action when necessary to appease domestic nationalists, it will try to do so in a way that minimizes the possibility for escalation—for example, by issuing statements instead of taking more strident action (Zhao, 2004, pp. 7–9, 15–16). This model is largely associated with Deng Xiaoping and other earlier Chinese leaders; it may not apply well to Xi's China. That said, Xi is still constrained by economic realities and might be under pressure from the party to avoid aggressive action that would jeopardize trade or cause other countries to balance against China (Blackwill and Campbell, 2016, pp. vii, 10, 28–29).

The **revisionist China** model sees Xi Jinping as very ambitious about lifting China's power and status regionally and in the world. Regionally, it seeks to establish China as unquestionably dominant in East Asia. Globally, it aspires to China being highly influential and respected—a model for countries worldwide (Schrader, 2020). Such a China may be driven by confidence in its own increasing ability to shape the behavior of other states, impatience with any challenge or attempts at containment from other states, and a belief that emphasizing peaceful relations over the assertion of Chinese rights will embolden Beijing's enemies (Zhang, 2013, pp. 106–110). It is also driven by a nationalistic desire to correct past injustices inflicted by Western powers and Japan during what China regards as the century of humiliation, from 1839 to 1949, by reclaiming China's rightful place at the center of Asian and world politics.[4] Although China will not want a large-scale war with the United States, it will be willing to risk such a war if necessary to protect its prerogatives as a great power.

---

[4] See Watson, 2019, pp. 16–20. Many authors argue that this sense of injured nationalism was cynically fostered by Chinese Communist Party officials who realized that, after Mao's failures and Deng's reforms, Marxism could not give the party the legitimacy it needed. Following this logic, such authors argue that the party is now the prisoner of its own creation—forced to adopt aggressive nationalist policies to appease a public that it has taught to resent foreign encroachment. It is also possible that many Chinese leaders may truly be animated by the same sense of aggrieved nationalism felt by the Chinese public and not merely constrained by it. Their ambitions may be strong (see Pillsbury, 2015, Chapter 1). See also Blackwill and Campbell, 2016, p. 14.

Finally, the **collapsing China** model envisions a China in which a slowing economy and faltering legitimacy have led to serious domestic challenges for the Communist Party of China. Under these conditions, the party may lash out abroad to project messages domestically, such as to encourage nationalism and justify clampdowns at home. Beijing may also lash out to secure markets or natural resources to bolster its faltering domestic economy. Such a China may engage in risky and aggressive action even if victory is unlikely (Beckley, 2019, p. 84).

Chinese analysts also have a number of different ways of viewing the United States. These overlapping perceptions of the United States (summarized in Table 4.2) exist independent of the relative aptness of the China models in Table 4.1.

Although informative, the models in Table 4.1 and Table 4.2 were not quite what we needed for systematic analysis and exercises. Thus, as discussed in the next section, we constructed alternative Reds and Reds' Blues that are structured more analytically *for the purposes of our prototype exercise.* As a group, the ideal set of Reds should cover adequately the range of behaviors suggested by the China-watcher Reds but also include other plausible Reds. By *cover*, we mean that, together, the Reds would assure thinking broadly enough to capture the possibilities most relevant to strategy choice—not in detail but in broad strokes.

Analytically Constructed Models of China

In constructing our alternative Reds, we used the template from the earlier chapter (see Table 3.3). Some 243 different models are possible within that structure, but Table 4.3 identifies three models of Red that we deemed adequate to cover the issues for our particular experimental application of method. Model A (cautious realist) represents a China that avoids risk but is otherwise rational and unbiased; it assumes that the United States will likewise work rationally to increase its power by competition (not war). Model B (accomplished gambler) predicts a more aggressive China willing to take risks and accept some losses to make significant gains. It sees the United States as eager to hold China down and as sometimes impulsive and risk taking. China Model C (contemptuous optimist) is also aggressive in seeking rapid

Table 4.2
Alternative Chinese Models of U.S. from China-Watcher Literature

Model	Motivations	Intentions Toward China	Willingness to Raise Tensions	Willingness to Risk Major Conflict
Election driven[a]	Winning elections	Mercurial—aggressive around elections or when needed to boost polling numbers, but more passive other times	High if U.S. prestige is at stake or, if need exists, to score points domestically	Low—after periods of bluster, United States will become preoccupied with other matters, especially economic
Classic defending hegemon[b]	Prevent rising challenger from displacing the United States as global hegemon	Containment—to prevent China's rise, undermine China's international position, build balancing coalitions, and sabotage its economy	Opportunistic—seize opportunities to weaken or damage China	Moderate—while preferring to avoid war, U.S. willingness to start a major regional war cannot be discounted. The United States may even see such war as inevitable
Impulsive president[c]	Personal power and fame	Mercurial—president will seek flashy wins by raising tensions as necessary or, in contrast, by slaughtering sacred cows to make deals	High	High—while not wanting a major conflict, the U.S. president may be willing to bluster, threaten, and accept high risks to seem strong and to shock his or her opponent and domestic audiences. Result could be unintended war
Aggressive imperialist[d]	World domination	Aggressive—the United States will not rest until the Chinese Communist Party (and any other organization that defies it) has been destroyed	High	Moderate

NOTES: Table is informed by a variety of sources, but the abstractions and labels are ours. The authors cited here seldom use precisely the words of our table and often have nuanced discussion of possibilities, rather than a clear assertion of U.S. behavior.
[a] United States as election driven (Wang Jin [王锦], 2020; Zhang Wenzong [张文宗], 2020a; Zhang Wenzong [张文宗], 2020b).
[b] United States as defending hegemon (China Taiwan Network [中国台湾网], 2014; Hu Jiping [胡继平], 2018).
[c] United States with an impulsive president (Chen Hongliang [程宏亮], 2020; Gong Zheng [龚正], 2020; Seeking Truth, 2019).
[d] United States with an impulsive president (Seeking Truth, 2019; Chen Hongliang [程宏亮], 2020; Gong Zheng [龚正], 2020).

Table 4.3
Alternative Models of China for Analysis

Model of China	Ambition/ Motivation (with respect to power)	Reasoning Style	Assessment Accuracy on Objective Matters	Baseline Model of the United States (can change with situation and history)
A. Cautious realist	Steady net progress	Rational-analytic and cautious	Accurate	RB1: firm, rational, competitive
B. Accomplished gambler	Rapid progress (and no losses along way)	Rational-analytic but forward-leaning and potentially emotional when losing (in domain of losses) or when seeing great opportunities	Accurate	RB2: intent on holding China down, impulsive, risk-taking
C. Disdainful optimist	Rapid progress (and no losses along way)	Rational-analytic but forward-leaning and potentially emotional when losing (in domain of losses) or when seeing great opportunity	Optimistic	RB3: whether or not rhetorically firm and blustery, ultimately weak and slow, rational but very cautious and lacking in resolve

NOTES: The possible choices (i.e., qualitative values) for the dimensions are as follows:
- Ambition/motivation: Steady net progress, steady net progress and no losses along way, rapid progress, rapid progress with no losses along way.
- Reasoning style: Rational-analytic, rational-analytic and cautious, rational-analytic forward-leaning, rational-analytic but forward leading and risk-taking in emotional circumstances, and rational-analytic but forward leaning and impulsive.
- Assessment accuracy: Conservative-pessimistic, accurate, optimistic.
- Models of Blue: RB1, RB2, RB (see Table 4.4).

gains of position and power, but, unlike Model B, it hopes to do so with minimal losses because of the United States being a paper tiger—perhaps issuing harsh statements and posturing but ultimately lacking resolve. In Model C's view, the United States is likely to back down.

These are analytically constructed, rather than based directly on the literature, but the cautious realist is akin to the stability-seeking model and the accomplished gambler is akin to the cautious realist.

The disdainful optimist has little authoritative evidence to support it, but there do exist jingoistic Chinese voices that are assertive, ambitious, and without fear of conflict because of the expectation that the United States would fold. Some articles discuss these perspectives (Wuthnow, 2020; Simkins, 2019).

As part of this activity, we developed alternative Red's Blues as in Table 4.4, which is similar but not identical in form to that presented in Figure 3.5.[5]

Modeling China's Limited Rationality

We assumed that China would, of course, consider a range of options, but that its evaluation of the options would reflect limited rationality. We came up with a number of candidate options that China might consider when contemplating an invasion of Taiwan or alternatives. We then reduced the list to the three (shown in Table 4.5) to simplify group discussion and the elicitation of subjective estimates.[6] For brevity, Table 4.5 does not include the baseline of "take no actions." In an exercise, however, participants would want to find influence actions that make that baseline more attractive. If diplomatic discussions

---

[5] The following sections continue with the alternatives laid out in Table 4.3 and Table 4.4. Two examples of other models were suggested in an exercise during the writing of this report. The first was that of a "beleaguered realist," with leadership that is naturally inclined to be of the cautious-realist variety but that is faced with a situation of very adverse trends and strong internal pressures for decisive action now, before it is too late. The second was that of an "inchoate alarmist," a China beset with power struggles among factions with contradictory views; the result might be inconsistent behavior, such as sometimes flailing and sometimes paralysis, over time during a crisis. These models of China might also have more alarmist images of the United States than those in Table 4.4. For example, they might see the United States as erratic and/or risk-taking.

[6] In more realistic exercises, the options might be more subtle, richer, or devious. Discussion could be sensitive. In an exercise conducted after the writing of this report, the options considered included various *gray-area actions*, actions involving aggressive but covert activities, and an option (with deception efforts) seeking a coup de main against Taiwan, one without lengthy mobilization and opportunity for U.S. response. We note again that the notional options in this report were the result of independent analysis and are not intended to represent the views or judgments of the Office of the Secretary of Defense.

Table 4.4
Alternative China's Models of the United States

Red's Blue	Motivation: Strength of Interests Seen	Reasoning Style	Resolve	Image of Red (can change with situation and history)
RB1. Rational competitor	High	Rational/cautious	High	United States sees China as competitor
RB2. Assertive enemy	Very high (sees vital national interests at stake)	Rational but impulsive and risk-taking in stress	High	United States sees China as an enemy, a threat to its power
RB3. Weak blusterer	Very high and blustery or moderate but blustery	Ultimately risk-averse	Low	United States sees China as an enemy, a threat to its power

Table 4.5
Chinese Composite Options Regarding Taiwan

Option	Political	Military	Economic	Objectives
1. Coercion without physical attack	Issue tough statements, including threats	Threaten and show force; possibly prepare for war	Threaten or employ economic warfare, such as halting Chinese tourism and punishing Taiwanese companies	Deter provocations by Taiwan and the United States
2. Limited military action	Threaten invasion while taking limited military action	Blockade ports; conduct limited missile attacks	Pursue economic warfare (as in option 1)	Deter and coerce Taiwan and the United States
3. Full-scale invasion	Declare necessary defensive action to secure part of China	Invade Taiwan	Pursue broad economic actions	Conquer and assimilate Taiwan; establish clear regional Chinese sovereignty

were possible, for example, the United States might remind Chinese leadership of how aggression would undercut or rule out many other actions on the regional or world stage, as well as opportunities for further economic growth.

Heuristic Chinese Factor-Tree Assessment of Options

A perfect storm would exist if China, in considering the factors of the factor tree, felt strongly about the factors (Table 4.6)—that is, its leadership felt *driven* to do something militarily and saw prospects as high (i.e., as very good). Here the factors are being evaluated subjectively on a scale of 0 to 10, with a score of 8–10 being very high.

For the full-scale invasion, Chinese leadership sees its factor tree as shown in Figure 4.2. When sketching the tree, thick arrows can be used to indicate relative emphasis. The overall score rating of the option can then be made, roughly, by simple eye-balling rather than any careful calculation.

Deliberate Chinese Evaluation of Options

Having completed the heuristic factor-tree assessment, we also conducted a more deliberate and intendedly rational-analytic analysis of how we thought the different Chinas would be affected by the alternative U.S. strategies. This employed the method described in Table 3.6. To encourage coherence, one person developed a strawman to be reviewed and adjusted. This way, the team estimated the most likely, best-case, and worst-case outcomes for each of the options of Table 4.5. Exercise participants attempted to think in the spirit of each model of China: Models A (cautious realist), B (accomplished gambler), and C (disdainful optimist). For each, they estimated the most-likely, best-case, and worst-case outcomes of the option using a scale from 0 (catastrophically bad) to 10 (superbly good). Estimates reflected initial instructions about the scenario. For example, the scenario specified that, by the time in question, China has the physical sealift capacity for invasion. It also specified that the United States was being stretched by other military operations with a citizenry wary of other adventures abroad.[7]

---

[7] The initiating scenario was tuned to establish a plausible starting point that would make attack of Taiwan relatively more attractive. In exercises with a different initiating scenario,

Table 4.6
The Perfect Storm

	Motivation	Legitimacy	Prospects for Success	Acceptability of Other Considerations	Net Evaluation (if motivation and prospects get more weight)
1. Limited military action	9	9	4	3	6
2. Full-scale invasion	9	7	7	5	8

Figure 4.2
Perfect Storm Graphic

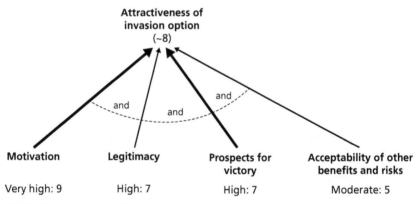

NOTES: The numbers indicate how highly the factors are scored subjectively on a 0 to 10 numerical scale. Thus, motivation is very strong and especially important. Legitimacy is less important except as a constraint (perceived legitimacy is necessary, according to the diagram).

Table 4.7 shows the results of our evaluation. The calculated results are shown in the grayed area, with bold, underlined characters indicating the option faring best for a given model of China. Such decisionmaking was more deliberate than the earlier heuristic evaluation but still ended up favoring attack for some models of China.

some of the models of Red would strongly prefer a strategy focused on subversion, political warfare, or various gray-area tactics.

Table 4.7
Chinese Deliberate Evaluation of Options

China Option	Outcome Model A: Cautious Realist			Outcome Model B: Accomplished Gambler			Outcome Model C: Disdainful Optimist			Model A's Net Outcome Assessment		Model B's Net Outcome Assessment		Model C's Net Outcome Assessment	
	Most Likely	Best Case	Worst Case	Most Likely	Best Case	Worst Case	Most Likely	Best Case	Worst Case	Weights (most likely: best-case: worst-base) for normal and emotion-heavy deliberations					
										Normal	Emotional	Normal	Emotional	Normal	Emotional
										2:1:3	2:1:3	2:2:1	2:2:0	1:2:0	2:2:0
1. Coerce without physical attack	3	4	2.8	3	3	2.8	3	4	2.8	3.1	3.1	3	3	3.4	3.5
2. Limited war	3	3	2.8	3	5	2.8	3	7	3	2.9	2.9	3.8	4	4.6	5
3. Full-scale invasion	3	8	0	5	9	3	7	9	5	2.3	2.3	6.2	7	7.4	8

NOTES: Calculations assume linear weighted sums with the weights indicated. Exercise participants estimate values on a 0 to 10 scale (extremely bad to extremely good) for the circumstances of the specified scenario. Bolded text is done for emphasis. Red text documents data from the exercise.

Both Models B and C evaluated the invasion option most favorably. Table 4.7 also distinguishes between estimates of Red's assessments for normal conditions and conditions in which emotions are high, whether related to nationalistic passions, fears of big strategic and political losses, anger, or something else. In this particular exercise, emotion did not play a major role in China's decision.

Figure 4.3 arrays the results from Table 4.7 as a chart. As expected, results vary substantially across models of Red. The cautious realist (Model A) favors the coerce-without-physical attack option. Models B and C strongly favor the invasion but for different reasons. The accomplished gambler (Model B) sees the invasion as a good bet, given the high upside potential, the potential for establishing once and for all Chinese domination over East Asia, and being rid of the Taiwan albatross. Model B sees considerable risk but gives the worst-case outcome less weight. In some circumstances (characterized as emotional), its enthusiasm for the best case might cause the

Figure 4.3
Contrasting Assessments for Alternative Models of Red

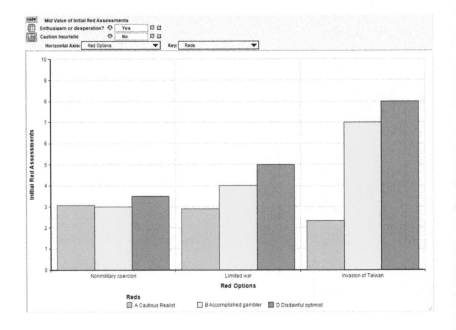

worst case to be ignored altogether. To the contemptuous optimist (Model C), it seems very likely that the United States (and perhaps Taiwan) will simply fold. It does not see much of a downside risk, other than possibly being wrong about the United States, a possibility on which Model C puts no weight if enthusiasm is running strong.[8]

Some might be surprised that the cautious realist appears to rate the options as comparable. This illustrates a shortcoming in standard mathematical methods in representing human reasoning. A real-world cautious realist or decisionmaker does not merely give less weight to bad cases. Often, such a decisionmaker applies a heuristic to the effect that if the outcome of the worst (plausible) case is bad enough, then the option should be discarded with prejudice without further thought. If we assumed such a heuristic, then the analogue to Figure 4.3 would be Figure 4.4. The primary effect, then, is for the cautious realist model of China.

Some Circumspection: Is This Adding Value?

We asked ourselves whether this deliberate analysis was accomplishing anything beyond what could be accomplished with text or a tabletop discussion. If so, it is for reasons related to the tyranny of the best estimate described in Chapter Two. Historically, positions that deviate markedly from the best estimate are often not given sufficient attention. Appointing a devil's advocate has sometimes been suggested, but the scholarly literature indicates that this method has often not fared well in practice.[9] Everyone knows who has been appointed to

---

[8] Japanese leaders had characteristics similar to Models B and C before launching the attack on Pearl Harbor (Hotta, 2013). Although knowing that objective prospects were poor, they acted as though, somehow, the best case would be accurate. They were affected by faith in Japanese moral superiority, their culture of fierce warfare, and the inability of the United States to stay the course. The analogy, however, is strained. In psychological terms, Japan was clearly in the domain of losses in 1941, whereas China is now more in the domain of gains.

[9] The usual example is that George Ball's pessimistic appraisals about Vietnam policy during Johnson administration meetings were ultimately ineffective. At the time, some argued that Ball was being used by Secretary of State Dean Rusk and President Johnson; he was seen a sophisticated yes man who could be relied on to raise cogent objections and then be overruled. Because of this, critics could not say dissenting views were unheard in policymaking circles

Figure 4.4
Contrasts If the Cautious Realist Is Even More Cautious

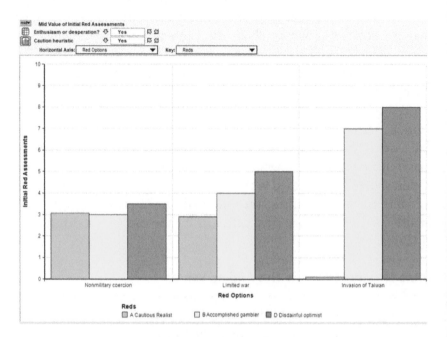

do the box-checking; they listen, nod their heads, and move on without really changing their thinking. This is especially true in informal decisionmaking. We would like to believe that formalizing the process of considering most-likely, best-case, and worst-case outcomes would encourage doing better. There is reason for some guarded optimism because some decisionmaking processes do in fact lead to better hedged decisions and avoidance of profound risks. For instance, venture capitalists do not mindlessly pursue options with great upside potential. They also consider downside risks (e.g., that a company's product would be prohibited by the government leading to bankruptcy). In March 2020, some governments contemplating what to do about COVID-19 acted because they were frightened by analysis,

---

(McFadden, 1994). Some discussion of Israel's use of the devil's advocate method appears in a paper diagnosing Israel's being surprised in the Yom Kippur War (Bar-Joseph and Kruglanski, 2003). Charles Schwenk discusses how the devil's advocate approach can be more effective in an informal review (Schwenk, 1984).

particularly for worst-case scenarios. It was difficult for them to ignore the potential for deaths running into the millions in the absence of strong intervention.[10]

Improved Procedures for Subsequent Exercises

This exercise merely illustrates the approach with a rough first-cut effort. In future exercises, these two alternative approaches would be more suitable:

- Individuals enter their estimates separately (either before or during the group session), and a control team tallies the results. If the estimates vary significantly, the control team reports them by perspective. For example, some participants might believe more than others that the invasion of Taiwan would be much easier. Because of this, results are reported for the two perspectives. This approach could be done in a half- or full-day seminar exercise without requiring high gaming skills of those conducting it.
- Participants break into smaller teams chosen so that participants with related attitudes are together. Teams might represent each of the three models of China (A, B, and C). Notes from the teams' discussions would probably be more coherent in conveying the reasoning of the alternative Chinese models. This approach might best be followed under the auspices of an organization skilled in creating and making use of parallel teams (e.g., war-college gaming experts).

---

[10] An influential report from Imperial College London (Ferguson, Laydon, Nedjati-Gilani, Imai, Ainslie, Baguelin, Bhatia, Boonyasiri, Cucunubá, Cuomo-Dannenburg, et al., 2020, p. 7) predicted 2.2 million deaths in the United States without mitigation. The White House drew heavily on more purely empirical modeling of the Institute for Health Metrics and Evaluation at the University of Washington.

Crafting U.S. Influence Strategies

Finding Targets for Influence

After developing a factor tree for Red, the Blue team agreed to immediately suggest a number of factors that Blue might reasonably attempt to affect favorably. In a traditional physical exercise, this might have meant using a marker to circle a drawing on a whiteboard. The electronic equivalent is shown in Figure 4.5. The items highlighted with asterisks are merely illustrative.

Candidate Actions for Influence

The team identified possible Blue influence actions (tactics) as shown in Table 4.8 using the method suggested in Chapter Three ("Influence Actions and Strategies" section). The influence tactics include two gray rows in Table 4.8, which represent the providing of what some in the Blue team considered as potential reassurances for China to negotiate and cooperate or actions dissuading China from favoring the invasion option. We did not pursue reassurances and other forms of nonmilitary dissuasion because doing so would have digressed too much from our Department of Defense–oriented exercise, but a number of possibilities (hawkish, dovish, and owlish) were raised and debated in offline work and virtual meetings. No consensus emerged, but the following serve as examples of possibilities:

1. Dissuading aggression. This option had in mind a broad series of actions in many domains, many of them political or economic, to counter China's efforts to marginalize Taiwan. These actions could include villainizing China to the world audience for its aggression toward Taiwan and ramping up military sales to and military cooperation between the United States and Taiwan.[11] The intent would be to dissuade China from believing that its more aggressive actions were fruitful.

---

[11] These actions would be consistent with other U.S. actions (Wong and Swanson, 2020) and with COVID-related efforts to bring Taiwan into the World Health Organization (Chen and Cohen, 2020). Many such actions have bipartisan support in the U.S. Congress, as illustrated by near-unanimous passage of the Taipei Act in 2019 (U.S. Senate, 2019).

Figure 4.5
Finding Targets for Influence Actions

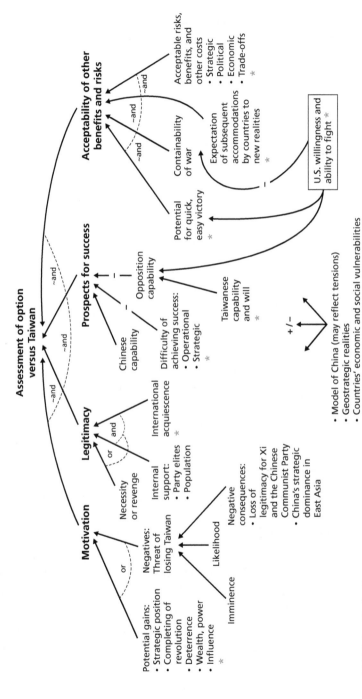

NOTE: * = illustrative.

Table 4.8
Candidate Actions for the United States in a Taiwan Crisis

Action Type	Action	Description
Political	Issue statements of concern	Not elaborated
	Recognize Taiwan	Recognize Taiwan as independent state
	Seek union with China that gives Taiwan a high degree of de facto independence	If of interest to Taiwanese (it is not of interest as of spring 2020), pursue negotiations with China for a mostly cosmetic union
	Seek Taiwan's independence and neutrality	Use UN mechanisms to elicit Taiwan's preferences and define a type of strict neutrality that would reassure China that Taiwan is not a base for U.S. aggression against China
Economic	Impose limited economic sanctions	Not elaborated
	Impose major economic sanctions	Not elaborated
Military	Pose show of force	Not elaborated
	Reinforce Taiwan	Not elaborated
	Supply Taiwan with advanced weapons	Not elaborated
	Deploy for possible war in Pacific	Not elaborated
	Prepare worldwide for possible war	Not elaborated

2. Union with China. If the Taiwan government was favorably inclined, propose China-Taiwanese negotiations with the intent of finding a one-China solution that would include special status for Taiwan and guarantees amounting to partial independence. The United States would promise to abide by any agreements between China and Taiwan. Some exercise participants immediately noted that Taiwanese nationals see themselves as independent and not part of China. After all, it has been more than 70 years since Chiang Kai-shek took his army to Taiwan and took control, still claiming to represent

China. With each subsequent generation, and particularly since martial law was lifted in 1987, people in Taiwan have come to see themselves more as Taiwanese than Chinese or even part of a greater China. As of 2019, 55 percent self-identified as Taiwanese only, with about 40 percent saying that they were both Taiwanese and Chinese.[12]

3. Independence but neutrality. Propose a UN-sanctioned referendum in Taiwan in which citizens would have two choices: (1) a one-China solution as noted in the previous option or (2) recognition of Taiwan as an independent state that would be rigorously neutral, accompanied by actions of both China and the United States to honor that neutrality. The formal expression of neutrality could be seen (by the United States at least) as reassuring China, although at some risk to Taiwan. Proponents argued that although China would probably initially hate and reject the idea of a UN referendum because it would anticipate that Taiwan would vote for independence, the concept of a Taiwan solution that would assure the island's neutrality might nonetheless be an attractive reassurance to those within China who see the Taiwan problem as poisoning prospects for many other things. Furthermore, proponents thought that, objectively, this would be a worthy approach. Others thought that the proposal would further antagonize China and make invasion more likely.

4. Tension-reducing changes of military forces. This envisaged an attempt to find some possible carrots that are attractive to China that would not be injurious to the United States and Taiwan. Some suggested an offer that, in exchange for China ceasing coercion of Taiwan and renouncing the use of force to bring about reunification, the United States would agree to some tension-reducing military actions, especially targeting those that feed into China's image of the United States seeing Taiwan as a

---

[12] See Su, 2019, and Zhong ChengFang [鍾辰芳], 2020. New polling data since the outbreak of COVID-19 suggest that the number of Taiwanese citizens self-identifying as Taiwanese only has grown considerably. For a news account, see Blanchard, 2020a.

convenient base for attacking China. Some suggested reducing the number of freedom-of-navigation operations that send Navy combatants through the Taiwan Strait. Others argued that such operations are especially important in resisting China's efforts to claim dominion in these international waters. Furthermore, the United States has long avoided provocative exercises near Taiwan and does not currently station military forces there. A joint command there was disbanded in 1979, after President Jimmy Carter established diplomatic relations with China under Deng Xiaoping. Thus, there may be little more for the United States to do. A different exercise might have suggested an option that would limit arms sales to Taiwan. That would also have been very controversial.

This simulated exercise illustrated how controversial ideas can usefully arise and be debated, perhaps opening minds to possibilities even if no consensus is reached on whether such policies would reassure China. Ultimately, any ideas would be worked out at the highest level of the U.S. government, quite possibly with disagreements among the U.S. Navy, the U.S. Department of Defense, the U.S. Department of State, and National Security Advisor.[13]

Influence Strategies

Using the tactics outlined in Table 4.8, we defined some alternative Blue influence strategies in Table 4.9. These are labeled in ascending order of what might be called strength of U.S. response, except that some of the options include the imagined incentive packages.

---

[13] Debate in our simulated exercise reminded us of how staff-level exercises can easily rule out options that leaders subsequently embrace. Examples include President Nixon's 1972 opening to China (the Shanghai Communiqué) and President Carter's 1979 recognition of the government of the People's Republic of China as the sole legal government of China. Both were controversial. More recently, and despite harsh words from President Xi, President Donald Trump has tightened relations with Taiwan. Also, the 2019 Taipei Act expressed strong U.S. support, including the statement "Taiwan is a free, democratic, and prosperous nation of 23,000,000 people and an important contributor to peace and stability around the world" (U.S. Senate, 2019). The admonition, then, is that exercises should not easily discard options.

Table 4.9
Influence Strategies for the United States in a Taiwan Crisis

Option	Political				Economic		Military				
	Issue Statements	Recognize Taiwan	Seek Unification Giving Taiwan Special Rights	Seek Taiwan's Independence and Neutrality	Impose Limited Sanctions	Impose Major Sanctions	Pose Show of Force	Reinforce Taiwan	Supply Missiles to Taiwan	Deploy for Pacific War	Prepare for World War
1. Weak	1	0	0	0	0	0	1	0	0	0	0
2. Modest	1	0	0	0	1	1	1	0	0	0	0
3. Moderate	1	0	0	0	0	1	1	1	0	1	
4. Strong equivocal	1	0	0	0	1	1	1	1	1	1	0
5. Strong	1	1	0	0	1	1	1	1	1	1	1
6. Strong, equivocal, variant 1[a]	1	1	0	1	1	1	1	1	1	1	0
7. Strong, equivocal, variant 2[a]	1	0	1	0	1	1	1	1	1	1	0

[a] Options 6 and 7 are alternative ways conceived to appear tough but provide opportunities for diplomacy. Only some participants believed that these would in fact induce diplomacy. Red text documents data from the exercise.

Estimating Effects of U.S. Strategy

We estimated the potential effectiveness of the Blue strategies in two ways—a fairly direct method and one involving more-deliberate calculations.

Direct Assessments of U.S. Options

Table 4.10 shows results from the Blue team's initial direct discussion of the likely effects of the various Blue influence strategies. This ad hoc assessment was done without structure or decision aids. As suggested earlier in Table 3.11, this assessment describes a consensus view of the group rather than distinguishing among Blue perspectives. The consensus view describes the Blue assessment of how the different China models might react to each of its strategies when contemplating invasion of Taiwan. The U.S. team concluded that Model A of China,

Table 4.10
China's Possible Reaction to Blue Actions When Assessing Its Own Invasion-of-Taiwan Option

	Model of China			
Blue Strategy	Model A Cautious Realist	Model B Accomplished Gambler	Model C Disdainful Optimist	Comments
1. Weak		+	++	
2. Modest		+		
3. Moderate		+		
4. Strong, equivocal		-	-	
5. Strong		+	+	Would cause Model C to rethink
6. Strong, plus incentives	--	+	+	Would cause Model A to see opportunity for resolution of long-standing problem; might cause Model C to rethink but take on character of Model B because of threat

NOTE: See option descriptions in Table 4.9. Red text documents data from the exercise. + indicates Red would be more likely to attack. - indicates the opposite.

the cautious realist, would not change its views when seeing any of the U.S. strategies, except that it might be intrigued by the possibility, under the last two options, of finally resolving the Taiwan problem permanently through negotiations. Model B of China, the accomplished gambler, would be more incentivized to invade Taiwan with any of the relatively weak U.S. strategies. It might pause and rethink when seeing a strong-equivocal U.S. reaction; the strength of the reaction would indicate resolve and the increased likelihood of war, while the failure to recognize Taiwan would indicate continued hard-headed rationality. In contrast, the unequivocally strong U.S. strategy would be even more alarming and might suggest that war was inevitable, something to be won. Model C of China, the disdainful optimist, would be shaken by a strong and unequivocal U.S. response because such a response would indicate a surprising resolve. It might then take on the character of Model B.[14]

More-Systematic Assessment
Heuristic Assessments Using Factor Trees
In a more systematic but still heuristic effort, we used the factor tree to estimate our notions of how effective the alternative U.S. strategies would be. Table 4.11 shows the results of our work through the entries for Models A, B, and C, and then combining them. It does not distinguish between perspectives because big differences of view did not arise in the discussion about this.

Deliberate Assessments Focused on Outcome Uncertainties
Earlier, we used the deliberate approach to estimate initial Chinese assessments of its alternatives. This is illustrated in Table 4.7, in which Blue's assessments are listed numerically from 0 (very bad) to 10 (very good). At this point, we asked ourselves whether the various U.S. strategies being considered would affect Chinese thinking about the most worrisome option—full-scale invasion of Taiwan. Table 4.12 shows our estimates for each model of China.

---

[14] It is possible to use computational modeling to suggest to a Red team how it might update its assessment of Blue—in this case, China's model of the United States. We illustrate this using Bayesian updating in Appendix C.

Table 4.11
Direct Estimates of Effects on China's Enthusiasm for Invasion Option

Factor	Motivation			Legitimacy			Prospects			Acceptability of Other Benefits and Risks			Net Effect (estimated directly)		
Model of Red	A	B	C	A	B	C	A	B	C	A	B	C	A	B	C
Blue Strategy	CR	AG	DO	CR	AG	DO	CR	AG	DO	CR	AG	DO	CR	AG	DO
1. Weak		+	+					+	+		+	+			
2. Modest		+	+					+	+		+	+			
3. Moderate	-	+	+				-	+	+	-	+	+		+	
4. Strong equivocal		-						-			-				
5. Strong		+	+					+	+		+	+			
6. Strong equivocal, variant 1		+	+					+	+ 1		+	+			
7. Strong equivocal, variant 2	--	+	+	--			-	+	+	-	+	+	-		

NOTES: CR = cautious realist; AG = accomplished gambler; DO = disdainful optimist.

Seeking a Robust U.S. Influence Strategy

As discussed in Chapter Three ("Seeking a Robust Strategy" section), we did not attempt to find a robust strategy analytically but instead relied on discussing as a group the relative merits of the strategies, noting the uncertainties. Our reasoning is as follows:

- The more-restrained strategies (strategies 1–3) in Table 4.10 would have no positive influence and might even be counterproductive. Some plausible Chinese reasoning would see such actions as a sign of U.S. weakness, not restraint.
- The strong strategy, which includes recognition of Taiwan, might do well but might also be counterproductive or even escalatory.

Table 4.12
Effects of U.S. Strategy on Chinese Assessment of Invasion Option

| Blue Strategy | Model of China | | | Comments |
	Model A Cautious Realist	Model B Accomplished Gambler	Model C Disdainful Optimist	
Original assessment of invasion option	2.5	3.9	8.3	
1. Weak	3	5	2	
2. Modest	3	5	2	
3. Moderate	3	5	2	
4. Strong, equivocal	2	5	7	Would cause Model C to rethink his model of the United States
5. Strong	2	4 or 6 (disagreement)	9	Would cause Model C to change image of the United States and behave more like Model B
6. Strong + incentives	1	2+	9+	Would cause Model A to see opportunity for resolution of long-standing problem. Might cause Model C to rethink but take on character of Model B because of threat

NOTE: Red text documents data from the exercise.

It might cause a forward-leaning China (the accomplished gambler) to be more alarmed about losing big. That would be troublesome, especially if China's leadership was under emotional stress. Although the contemptuous optimist would be shocked into rethinking, they would be more likely to revert to Model B behavior than not because loss of Taiwan would be offensive.

- The potential problems with the strong strategy would not be mitigated by the incentives, although it might be to the cautious realist China.
- A hybrid is needed (new last row in Table 4.13).

Table 4.13
Revised Option Assessment

	Model of China			
Blue Strategy	Model A (Cautious Realist)	Model B (Accomplished Gambler)	Model C (Disdainful Optimist)	Comments
1. Weak		+	++	
2. Modest		+		
3. Moderate		+		
4. Strong, equivocal		-	-	
5. Strong		+	+	Would cause Model C to rethink
6. Strong + incentives	--		-	Would cause Model A to see opportunity for resolution of long-standing problem; might cause Model C to rethink but take on character of Model B because of threat
7. Hybrid: strong but equivocal + incentives	-	+/-	+/-	Model A would see opportunity; Model B and C would be similar and might or might not change their view

NOTE: + indicates Red would be more likely to attack.

The most robust influence strategy would be a strategy not on the original list. It would be strong but equivocal, perhaps with incentives or perhaps even more provocative). To be more precise, unless the United States saw virtue in a war that might be bloody but ultimately successful for it (i.e., to the extent that the U.S. objective was deterrence of China and a possible resolution of issues), then the hybrid strategy would be superior. This was all quite speculative and beyond the pay grade of exercise participants. Thus, the exercise served merely to open minds, raise possibilities, and sharpen differences of view.

As noted earlier, this chapter describes our simulation of an exercise but not a true exercise with a fresh set of U.S. participants. In future work, we anticipate running multiple exercises, potentially with different constructions of Red, Blue influence strategies, and conclusions about the way ahead. Ideally, the cumulative insights from multiple exercises, supported by more careful in-depth analysis, would prove useful in preparing for possible real-world crisis. As we learned in recent months, virtual exercises are feasible but are not as effective.

CHAPTER FIVE

Conclusions

In this report, we described the motivation for and elements of an approach to using wargames and other human exercises to aid strategy development when dealing with an adversary whose perceptions and reasoning are uncertain. We also illustrated the approach with a simulated exercise examining a crisis involving China and Taiwan. The application is purely notional and fairly abstract (much more so than would be possible in an inside-the-government exercise). We should emphasize that the approach imagines circumstances that do not currently exist and that such a crisis may never arise. Our intention was to use the cognitive modeling approach to understand how an adversary might be reasoning *if* it contemplated a highly aggressive action (in this case, an invasion) and how alternative U.S. strategies might or might not be influential in avoiding that aggression. With respect to China in the example, rather than embrace a worst-case analysis, we ask about the circumstances in which a future China might plausibly contemplate invasion of Taiwan and what factors would influence China's judgments. Some factors might include influence by the United States, deterrent capabilities, and coercive diplomacy. Others might be tied to internal Chinese politics out of reach of U.S. influence efforts. Even so, a good U.S. influence strategy might favorably influence Chinese perceptions about the need to act aggressively, the legitimacy of doing so, prospects for success, and both short- and long-term risks. Even if influence strategy did not change China's best estimate of prospects, it might greatly increase China's worries about the risks.

Our work in this report was an early prototype effort. We believe that the ideas and methods are now ready for more-realistic experimentation, including experiments within the U.S. government. Such experiments will be valuable in themselves and will also help improve the methods. Tentatively, we see value in exercises for cases in groups identified in Table 5.1. These include applications to (1) different potential adversaries, (2) long-term strategic competition, (3) gray-zone activities, and (3) crises at different levels of potential violence (e.g., low-level war with the potential of escalation). Each such application will require an in-depth study of substantive issues and a refinement of methods and tools, but the potential scope of applications is wide.

Table 5.1
Topics for Possible Exercises

Group	Adversary	Type Issue	Locus	Level(s)
1	China	Deterrence	Taiwan	Blockades, missile exchanges, invasion (including a short-mobilization variant)
2	China	Escalation	Taiwan	Blockades, missile exchanges, invasion variants, broader war with the United States, nuclear war
3	China	Deterrence and coercive diplomacy	South China Sea	Creeping aggression
4	China	Escalation	South China Sea	Events leading to shooting exchanges or war
5	China	Strategic competition	Asia-Pacific	Force postures, alliances, strategic maneuvers
6	Russia	Deterrence	Baltics	Gray zone, hybrid warfare, invasion
7	Russia	Deterrence	Ukraine	Gray zone, hybrid warfare, invasion
8	Russia	Strategic competition	Europe	Force postures, alliances, strategic maneuvers

Appendixes

The following appendixes provide additional detail on some aspects of
the report:

- A. The mathematics of the thresholded linear weighted sum
  formula referred to in the report.
- B. An overview of a computational model that we developed to
  supplement the work of the main report.
- C. A model illustrating use of Bayesian updating as a possible
  decision aid for teams in an exercise.
- D. Some detailed but altogether notional data for use in the
  computational model for China.
- E. Approximating uncertainty by referring to best-estimate, best,
  and worst cases.

Mathematics of Thresholded Linear Weighted Sums

The method of thresholded linear weighted sums was introduced in earlier work (Davis and Dreyer, 2009; Davis and O'Mahony, 2013). It is a simple and practical way to approximate the evaluation of a nonlinear function characterizing a system with a number of critical components—i.e., components that must *all* work if the system as a whole is to work. We specify requirements (i.e., minimum values for the effectiveness of each component). Then, if all the requirements are achieved, the system works, and we assume that the overall effectiveness increases linearly with improvements in any of the components. Some increase more rapidly than others, as reflected by weighting factors.

For the context of this report, suppose that Red is assessing an option. For the option to look favorable, it must pass tests of Motivation (*M*), Legitimacy (*L*), Prospects for Success (*P*), and Acceptability (*A*) of other costs and risks. *F* is the set of factors, and F_i is the *i*th member of the set. We assign scores to each factor on a 0 to 10 scale and assess the option's attractiveness score *S* using a simple formula:

$$S = \begin{cases} 0 \text{ if } F_i < F_i^0 \text{ for any } i \\ \hline \displaystyle\sum_{i=1}^{4} W_i F_i \text{ otherwise} \end{cases}$$

F_i^0 = threshold for factor *i*

W_i = weight for factor *i*.

This can be expressed more compactly as

$$\text{If } Min(F_i^0 - F_i, F_i \in F) < 0$$
$$\text{Then } 0$$
$$\text{Else } W \cdot F,$$

where the last item refers to the scalar or dot product of W and F.

Computational Version of Factor Tree Model

This appendix provides an overview of the first version of a computational model we constructed that corresponds to the test case in Chapter Four. The model describes a simulated exercise to address China posing a threat to Taiwan. The model uses alternative models of China's reasoning and inputs estimates of the effectiveness of U.S. influence strategies for those alternative models. It is not intended for real-time use during human exercises because computer models, even simple ones, tend to be distracting and can undercut the value of human play. However, such a model can be very useful for background analysis, especially the exploratory analysis needed to consider effects of uncertainty. In practice, such a model might be used iteratively. Expert deliberation would be performed "live" to interpret, agree on, and estimate values of model parameters, which would then be systematically explored with the computational model. Model results would then be presented again for additional insights and challenges. This process might continue through multiple cycles. The process parallels deliberation with analysis (National Research Council of the National Academies, 2009).

Figure B.1 shows the top-level user interface for the model. The user can first select one of two model modes from a computation method dropdown (not shown). The choices are deliberate assessment or factor tree assessment. The deliberate method uses expert inputs estimating how China (represented by alternative cognitive models) would judge outcomes of its several options as described in Chapter Four ("More Systematic Assessment" section). The factor-

Figure B.1
Model Interface

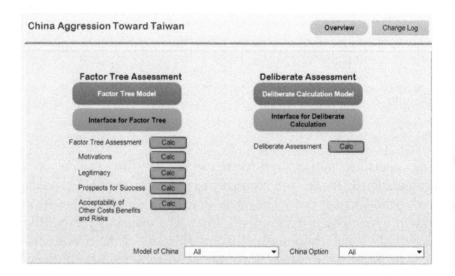

tree method is also illustrated in that chapter. In the model, expert judgments are used for variables, such as Motivation and Legitimacy.

The deliberate computation method (right side of Figure B.1) represents an intendedly rational-analytic assessment by China of several options under various conditions across both alternative China cognitive models and different U.S. strategies. Clicking the Interface for Deliberate Calculation button opens an interface (Figure B.2), where the user can input values from menus (determined in a separate deliberative exercise) into several tables, which are actually n-dimensional arrays. Outcome Estimates (in first column of Figure B.2) produces, for each model of China, China's estimate for three illustrative options (nonmilitary coercion, limited war, or invasion) of outcomes for most-likely, best, and worst cases. The calculation can also take into account whether China knows the U.S. strategy, in which case results are generated from Outcome Estimates by Blue strategy, a table identical to the above but with an additional dimension for U.S. strategy (weak, strong, strong equivocal). Weights (Outcome Weights Index) is a table defining weights placed

Figure B.2
Deliberate Calculation Configuration

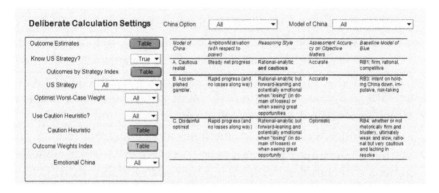

on each outcome condition for each model of China and for whether China is in an emotionally heightened state.

The factor tree computation method (left side of Figure B.1) method allows for multi-resolution input of factors contributing to China's assessment of an option. For each layer of factors, the user can either set the value of each factor directly or aggregate the value from lower-level factors. If any of the top-level factors (motivations, perceived legitimacy, prospects for success, net acceptability of other costs and risks) is set to not compute from subfactors, the user provides values for each option and each China cognitive model. Otherwise, factors are aggregated by using either thresholded linear weighted sums (see Appendix A) or a method that takes the value of the largest subfactor and adjusts it marginally based on the other subfactors.

The logic for both the factor tree and deliberate calculation can be examined by clicking on either model (see buttons in Figure B.1).

Note that most parameter inputs are from menus from which any of several discrete parameter values, or all can be chosen. If all is chosen, then the model runs for all of the discrete values with the results stored in a database so that the user can look at output displays, change parameter values from menus, and instantaneously see the relevant results. This allows for interactive exploration of what-ifs and, with more-elaborate manipulations, answers to such questions as "under what assumptions are the results good or bad?"

Figure B.3 illustrates use of the model for exploratory analysis in the factor tree computation method. In this case, the model is being used in its lowest-resolution form to assess the attractiveness of an especially aggressive option. One inputs the estimated values of Red's motivation, sense of legitimacy, prospects for quick and easy success, and acceptability of other considerations. Even if legitimacy is deemed low (3), as shown at the top of Figure B.4, as is the acceptability of other costs and risks, the option still looks attractive if motivation is very high, especially if prospects for quick success are also very high.

The model was based on methods analogous to those in an earlier model dealing with public support of terrorism. That model is documented in considerable detail in Davis and O'Mahony, 2013. The model was developed in Lumina's Analytica, a high-level programming language.

Figure B.3
Factor Tree Configuration

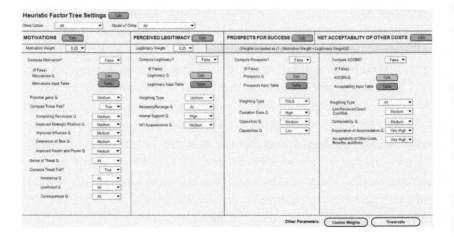

Figure B.4
Illustrative Exploratory Analysis

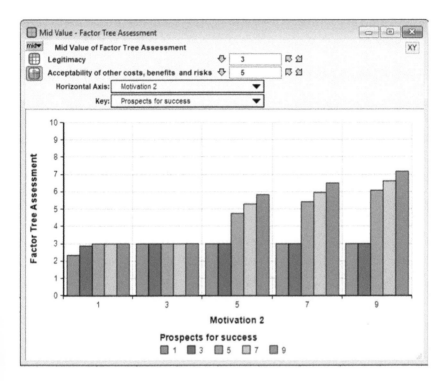

Bayesian Updating of Red's Blue: Simple Artificial Intelligence

One aspect of influencing the adversary is influencing how the adversary thinks about the defender—about the defender's interests and about the credibility of its deterrent threats. In this appendix, we describe a simple Bayesian computational model that we constructed to illustrate issues and dynamics. Such a model could be used as a decision aid in exercises. However, our example also demonstrates why dealing with such issues in a computational model is fraught with difficulties and apt to bury crucial assumptions if it is encoded in common programming languages.

The essence of the model can be understood from Figure C.1. The inputs include (left side) Blue's strategy, Blue's model of Red (Red), Red's initial model of Blue, and a likelihood function describing the subjective probability of Red's Blue taking various actions. The model represents Red's intelligent reasoning. Red has prior notions about Blue behavior, then observes Blue's actions. Red then reasons about whether Blue's actions suggest changing its perception of Blue. Are the actions consistent with what the initial Red's Blue might do, or are they more consistent with a different image of Blue? This reasoning is accomplished using Bayesian theory and produces an updated image of Blue. In such a Bayesian treatment, Red assigns probabilities to the alternative models of Blue. The updating adjusts those probabilities separately for each Blue action observed. Then the model considers all of the actions and decides whether to adjust its best-estimate characterization of Blue. In doing so, the model can treat some actions as more or less

Figure C.1
Bayesian Model of Red Updating Red's Blue

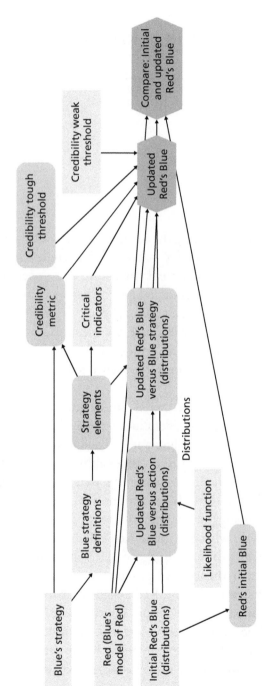

significant than others and can apply threshold criteria before changing models of Blue.

In a human game, this might correspond to the Red team observing Blue's actions and saying:

> Obviously, we have been misunderstanding with whom we are dealing. We have been going along with the best estimate that Blue is rational and cautious, eager to avoid war and unwilling to take overly provocative actions. Blue, however, has just raised stakes markedly by its diplomatic actions and by taking military actions preparing for full-scale war, actions that are potentially escalatory. We recognized that Blue might be willing to bluster, show force, and threaten in the nature of a paper tiger, but Blue's actions go beyond what we would expect from that behavior. Moving forward, we should assume that Blue has his back up, sees vital interests at stake, is determined, and is willing to risk escalation.

Or, conversely, it might conclude:

> We have been overestimating Blue. We anticipated that Blue would take strong diplomatic, economic, and military actions. We might not know how far he would go ultimately, but he would either be preparing for war or overtly threatening us even if he were secretly contemplating giving in. What we are seeing is a surprisingly timid response. Moving forward, we should assume that—even if Blue now takes some belated actions in an attempt to deter us—they will be the hollow threats of a paper tiger.

Although the basic concept is simple, implementing the model with a mathematical algorithm or computer model requires an apparatus allowing treatment of subjective probabilities and Bayesian updating. Even such a simple model depends on many inputs. Figure C.2 shows Red's Bayesian priors for each of the Red's models of

Figure C.2
Red's Bayesian Priors About Blue

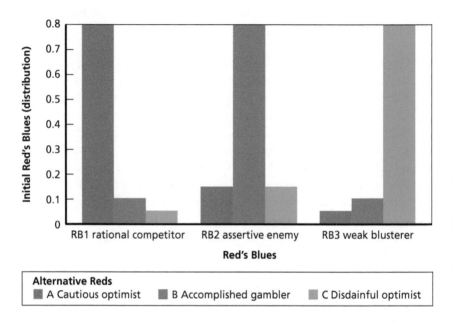

Blue. It uses Reds and Red's Blues in a way that is similar to those of Chapter Four, but Figure C.2 uses generic actions.[1]

Figure C.3 shows the Bayesian likelihood function used by each Red's Blue. For example, if Red's Blue is the rational competitor, then it will certainly issue diplomatic statements (probability of 1) in response to Red's actions but will be extremely unlikely (probability of 0.1) to prepare for global war. In another example from the third column, if Blue is a weak blusterer, many actions are reasonably probable but

---

[1] The difficulty in developing accurate likelihood tables such as this cannot easily be overestimated. Adversaries frequently misunderstand each others' perceptions, likely reaction to their own initiatives, and the reasons for those reactions. See, for example, Jervis, 2017 (Chapter 9), which includes discussion of how resistant leaders are to changing their beliefs. For a convinced leader, "only the most dramatic events will shake him," a point that Jervis credits to Glenn Snyder and Paul Diesing. In developing such estimated likelihood tables, analysts should be very skeptical about the effectiveness of usual diplomatic or even military signals in changing perceptions.

Figure C.3
Likelihood Function

	RB1 Rational competitor	RB2 Assertive enemy	RB3 Weak blusterer
Diplomatic statements	1	1	1
Strong international actions	0.05	1	0.05
Limited sanctions	0.9	1	1
Major sanctions	0.1	1	0.4
Show of force	0.8	1	1
Reinforcement of ally	0.5	1	0.25
Resupply of ally	0.1	1	0.5
Preparation for local war	0.2	1	0.2
Preparation for global war	0.1	0.5	0.05

Edit Table of Likelihood Function — Blue Actions ▼ — Red's Blues ▼

not strong on international actions,[2] preparation for local war, and especially preparation for global war (probability 0.05).

Figure C.4 illustrates one of the model outputs. It shows the updated probability distributions for different Red's Blues as a function of Blue's strategy (top, chosen here to be the strong strategy), Blue's model of Red (top), and the particular actions called for in Blue's strategy (the different colored bars, corresponding to the leftmost column of Table C.1.

The example is for the disdainful optimist (Model C of Red). It believes initially that Blue is very likely (probability of 0.8) to be the weak blusterer. It looks at each of Blue's actions under Blue's strong strategy and updates its assessment of Blue with just that action in mind. Most of the actions have little effect (the probability that Blue is the weak blusterer remains high, in the range 0.55–0.8), except for the strong diplomatic statements, preparation for global war, and—to a lesser extent—preparation for local war. If Red sees those actions, it will see a significant probability that it is facing RB2, the assertive enemy.

Recognizing likely demands by model users for a best estimate of Blue's Red, rather than a probabilistic statement across Blue's Reds, the model considers all the Blue actions and estimates the most likely

---

[2] "Strong international actions" might include evicting the adversary from international forums, recognizing enemies of the adversary, or preventing adversary use of the international financial system.

Figure C.4
Illustrative Model Results: Updated Red's Blue (Probabilistic)

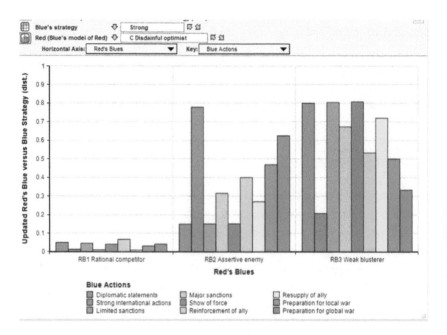

Red's Blue. That estimate depends on model details, such as how indicators are added up and what thresholds are used for shifts of Red's Blue. National intelligence organizations have complex methods for making such judgments, and their experts take into account numerous aspects of context unlikely to be present in a simple computational model. Even a Red team in a wargame will often try to do so. Thus, the purpose of presenting the model here is to merely illustrate what can be done computationally, while noting that many details need to be specified. It would be important, in considering an analogous model as a possible decision-aid for a human exercise, to ensure that appropriate people reviewed model details before using it. With a high-level language, such as the one we used, entering such human data is straightforward and making changes to better reflect context does not require professional programming skills. Nonetheless, the modeling is not trivial. One reason is that experts and policymakers often disagree about the significance of the various indicators. To illustrate how the disagreements can be dealt with, Figure C.5 compares the initial and updated estimates of Red's Blue as a function of perspective (the perspective of Blue analysts speculating about the importance of various Blue actions to Red's judgments about the nature of Blue).

In a wargame or other exercise, the Blue team might go through analogous reasoning about Red's perceptions, but it would do so much less systematically and without the benefit of definitions and structure. A computational model can augment human exercises by incorporating insights and estimating the effects of changed results for cases not discussed in the exercises.

The details of the model are best understood by looking directly at the self-documenting code, which is written in a visual-programming language (Lumina's Analytica®).

Figure C.5
Illustrative Model Results: Updated Red's Blue

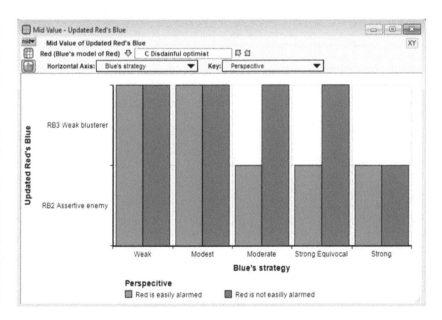

Input Data for Model

The deliberate calculation of the computational model described in Appendix B requires detailed subjective expert inputs. Here, we provide our detailed but altogether notional data for use in the computational model for China. Repeating Table 4.7 from the main text, we have Table D.1.

Table 4.12 indicated possible revisions of Chinese estimates as a function of what it observed as U.S. response to initial actions. Table D.2 is the result of combining the first three rows of that table and considering them as weak U.S. responses. Table D.3 is a more detailed interpretation of these earlier data, filling in data for other Red strategies. It assumes that the Chinese originally anticipated a strong but equivocal U.S. response.

Table D.1
Initial Chinese Assessment of Their Options

China Option Case	Outcome Model A: Cautious Realist			Outcome Model B: Accomplished Gambler		Outcome Model C: Disdainful Optimist			Model A's Net Outcome Assessment		Model B's Net Outcome Assessment		Model C's Net Outcome Assessment	
	Most Likely Case	Best Case	Worst Case	Most Likely Case	Worst Case	Most Likely Case	Best Case	Worst Case	Normal	Emotional	Normal	Emotional	Normal	Emotional
									2:1:3	2:1:3	2:2:1	2:2:0	1:2:0	2:2:0
1. Coerce without physical attack	3	4	2.8	3	2.8	3	4	2.8	3.1	3.1	3	3	3.4	3.5
2. Limited war	3	3	2.8	3	2.8	3	7	3	2.9	2.9	3.8	4	4.6	5
3. Full-scale invasion	3	8	0	5	3	3	9	5	2.3	2.3	6.2	7	7.4	8

Weights (most-likely: best-case: worst-case) for normal and emotion-heavy deliberations

Table D.2
Estimated Revisions When U.S. Strategy Is Observed

Blue Strategy	Model of China			Comments
	Model A Cautious Realist	Model B Accomplished Gambler	Model C Disdainful Optimist	
Original assessment of invasion option	2.5	3.9	8.3	
Weak	3	5	10	
Strong, equivocal	2	5	7	Would cause Model C to rethink model of the United States
Strong	2	4 or 6 (disagreement)	9	Would cause Model C to change image of U.S. and behave more like Model B
Strong + incentives	1	2+	9+	Would cause Model A to see opportunity for resolution of long-standing problem Might cause Model C to rethink but take on character of Model B because of threat

Table D.3
Detailed Subjective Estimates

Blue Strategy	Red Strategy	Outcome Model A Cautious Realist			Outcome Model B Accomplished Gambler			Outcome Model C Disdainful Optimist		
		Most Likely	Best Case	Worst Case	Most Likely	Best Case	Worst Case	Most Likely	Best Case	Worst Case
Weak	Coerce	3	4	2.8	3	3	2.8	3	4	2.8
Weak	Limited war	3	3	2.8	3	5	2.8	3	7	3
Weak	Invasion	3	8	0	5	9	3+2	7+1	9+1	5+2
Strong, equivocal	Coerce	3	4	2.8	3	3	2.8	3	4	2.8
Strong, equivocal	Limited war	3	3.1	2.8	3	5	2.8	3	7	3
Strong, equivocal	Invasion	3	8	0	5	9	3	7	9	5
Strong	Coerce	3	4	2.8	3	3	2.8	3	4	2.8
Strong	Limited war	3	3.1	2.8	3	5	2.8	3	7	3
Strong	Invasion	3	8	0	5	7	1	7-2	9-2	5-4

Simplifying Probabilistic Calculations

When contemplating a military option, a country's military staff might attempt to characterize the relative probability of different outcomes. The staff might generate a large number of scenarios and work their way through them with gaming or computer models that are adequate to estimate the quality of outcome. The staff might then construct a graphic showing the distribution across outcomes (Figure E.1). This would not truly be a probability distribution, but it might be discussed as such by associating probability with the fraction of scenarios in which an outcome was observed. When briefing the results, one might say, "Well, it's always possible that things will escalate to general nuclear war, which would be catastrophic [score of 0]. That happened in about 10 percent of cases. More typically, outcome was good [score of 7], and, in about 20 percent of the cases, outcome was spectacularly good [score of 10]."

Upon studying such a result, decisionmakers (or senior staff) might well conclude that they were going to ignore the true worst case (catastrophe) because it is too unlikely ("10 percent is like 0") to take seriously. They might then look at Figure E.1 and conclude that, for practical purposes, the most-likely, best-case, and worst-case outcomes should be considered to be 7, 10, and 5. Thus, they would characterize the analysis as shown in Figure E.2.

Characterizing the outcome by the weighted-sum formula would not be a bad approximation in this case if the weights used were 2:1:1 for most-likely, best, and worst cases, respectively. More generally, the shape of the distribution might be very different, in which case the

Figure E.1
Outcome Distribution

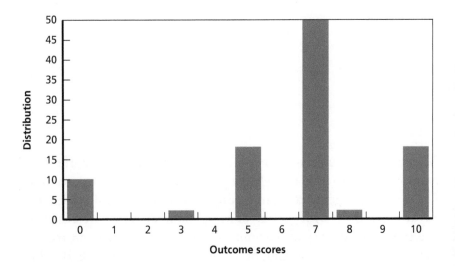

Figure E.2
Characterization After Thresholding for Plausibility

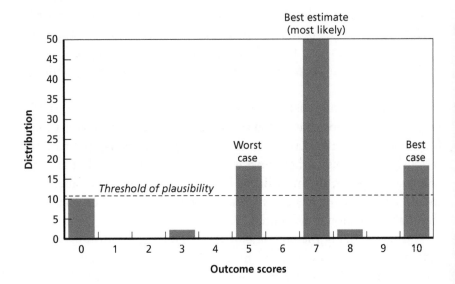

decisionmaker might be much more interested in the likelihood of the worst case—which would be very difficult to estimate with analytical integrity because the various scenarios are not equally probable. Judgments will be necessary.

References

Arquilla, John, and Maria Moyano Rasmussen (2001), "The Origins of the South Atlantic War," *Journal of Latin American Studies*, Vol., 33, No. 3, pp. 739–775.

Baker, Peter (2019), "Lesson on the Soviets and Afghanistan, Laid Out in a Cable," *New York Times*, Section A, p. 5.

Bar-Joseph, Uri, and Arie W. Kruglanski (2003), "Intelligence Failure and Need for Cognitive Closure: On the Psychology of the Yom Kippur Surprise," *Political Psychology*, Vol. 24, No. 1, pp. 75–99.

Beckley, Michael (2019), "The United States Should Fear a Faltering China," *Foreign Affairs*. As of October 7, 2020:
https://www.foreignaffairs.com/articles/china/2019-10-28/united-states-should-fear-faltering-china

Blackwill, Robert D., and Kurt M. Campbell (2016), *Xi Jinping on the Global Stage: Chinese Foreign Policy Under a Powerful but Exposed Leader*, New York: Council on Foreign Relations, Council Special Report No. 74. As of October 13, 2020:
https://cdn.cfr.org/sites/default/files/pdf/2016/02/CSR74_Blackwill_Campbell_Xi_Jinping.pdf

Blanchard, Ben (2020a), "In Taiwan, Anger at China over Virus Drives Identity Debate," Reuters. As of November 5, 2020:
https://www.reuters.com/article/us-health-coronavirus-taiwan-china/in-taiwan-anger-at-china-over-virus-drives-identity-debate-idUSKBN21J4JK

Blanchard, Ben (2020b), "Taiwan President Says Drills Show China Is Threat to Region," Reuters, Aerospace and Defense. As of November 11, 2020:
https://www.reuters.com/article/us-taiwan-china/taiwan-president-says-drills-show-china-is-threat-to-region-idUSKCN26B06I

Blanton, Tom, and Svetlana Savranskaya, eds. (2019), "The Soviet Invasion of Afghanistan, 1979: Not Trump's Terrorists, Nor Zbig's Warm Water Ports," George Washington University National Security Archive webpage, briefing book #657. As of April 17, 2020:
https://nsarchive.gwu.edu/briefing-book/afghanistan-russia-programs/2019-01-29/soviet-invasion-afghanistan-1979-not-trumps-terrorists-nor-zbigs-warm-water-ports

Bracken, Paul, Ian Bremer, and David Gordon, eds. (2008), *Managing Strategic Surprise: Lessons from Risk Management and Risk Assessment*, New York: Cambridge University Press.

Chen Hongliang [程宏亮] (2020), "Trump Is Angry! [特朗普气坏了!]," China Institutes of Contemporary International Relations webpage. As of August 11, 2020:
http://www.cicir.ac.cn/NEW/opinion.html?id=2825fdba-299b-4bcd-bdf4-df162b0f7eae

Chen, Yu-Jie, and Jerome A. Cohen (2020), "Why Does the WHO Exclude Taiwan," Council on Foreign Relations webpage. As of May 5, 2020:
https://www.cfr.org/in-brief/why-does-who-exclude-taiwan

Chernyaev, Anatoly S. (2020), "The 'Irreplaceable' Chernyaev Diary 1980," trans. Anna Melyakova, Svetlana Savranskaya, ed., George Washington University National Security Archive webpage, briefing book #706. As of May 25, 2020:
https://nsarchive.gwu.edu/briefing-book/russia-programs/2020-05-25/irreplaceable-chernyaev-diary-1980

China Taiwan Network [中国台湾网] (2014), "America Has Become a Paper Tiger, Who Is at Fault? [美国变成 "纸老虎", 谁之过?]." As of August 11, 2020:
http://www.taiwan.cn/plzhx/gjshd/201407/t20140717_6631596.htm

Copeland, Dale (2019), "Grappling with the Rise of China: A New Model for Thinking About Sino-American Relations," in Nicole Peterson, ed., *Chinese Strategic Intentions: A Deep Dive into China's Worldwide Activities*, Boston, Mass.: NSI Inc., pp. 1–6. As of October 13, 2020:
https://nsiteam.com/social/wp-content/uploads/2019/10/SMA-Chinese-Strategic-Intentions-White-Paper-FINAL-01-Nov-2.pdf

Craig, Campbell, and Sergey Radchenko (2018), "MAD, Not Marx: Khrushchev and the Nuclear Revolution," *Journal of Strategic Studies*, Vol. 41, Nos. 1–2, pp. 208–233.

Davis, Paul K. (1994), "Institutionalizing Planning for Adaptiveness," in *New Challenges in Defense Planning: Rethinking How Much Is Enough*, in Paul K. Davis, ed., Santa Monica, Calif.: RAND Corporation, MR-400-RC, pp. 73–100. As of October 13, 2020:
https://www.rand.org/pubs/monograph_reports/MR400.html

Davis, Paul K. (2003a), "Exploratory Analysis and Implications for Modeling," in Stuart Johnson, Martin Libicki, and Gregory Treverton, eds., *New Challenges, New Tools for Defense Decisionmaking*, Santa Monica, Calif.: RAND Corporation, MR-1576-RC, pp. 255–283. As of October 13, 2020: https://www.rand.org/pubs/monograph_reports/MR1576.html

Davis, Paul K. (2003b), "Uncertainty Sensitive Planning," in Stuart Johnson, Martin Libicki, and Gregory Treverton, eds., *New Challenges, New Tools for Defense Decisionmaking*, Santa Monica, Calif.: RAND Corporation, MR-1576-RC, pp. 131–155. As of October 13, 2020: https://www.rand.org/pubs/monograph_reports/MR1576.html

Davis, Paul K., and John Arquilla (1991a), *Deterring or Coercing Opponents in Crisis: Lessons from the War with Saddam Hussein*, Santa Monica, Calif.: RAND Corporation, R-4111-JS. As of October 13, 2020: https://www.rand.org/pubs/reports/R4111.html

Davis, Paul K., and John Arquilla (1991b), *Thinking About Opponent Behavior in Crisis and Conflict: A Generic Model for Analysis and Group Discussion*, Santa Monica, Calif.: RAND Corporation, N-3322-IS. As of October 13, 2020: https://www.rand.org/pubs/notes/N3322.html

Davis, Paul K., and Kim Cragin, eds. (2009), *Social Science for Counterterrorism: Putting the Pieces Together*, Santa Monica, Calif.: RAND Corporation, MG-849-OSD. As of October 13, 2020: http://www.rand.org/pubs/monographs/MG849.html

Davis, Paul K., and Paul Dreyer (2009), *RAND's Portfolio Analysis Tool (PAT): Theory, Methods, and Reference Manual*, Santa Monica, Calif.: RAND Corporation, TR-756-OSD. As of October 13, 2020: https://www.rand.org/pubs/technical_reports/TR756.html

Davis, Paul K., David C. Gompert, Stuart Johnson, and Duncan Long (2008), *Developing Resource-Informed Strategic Assessments and Recommendations*, Santa Monica, Calif.: RAND Corporation, MG-703-JS. As of October 13, 2020: http://www.rand.org/pubs/monographs/MG703.html

Davis, Paul K., Jonathan Kulick, and Michael Egner (2005), *Implications of Modern Decision Science for Military Decision Support Systems*, Santa Monica, Calif.: RAND Corporation, MG-360-AF. As of October 13, 2020: http://www.rand.org/pubs/monographs/MG360.html

Davis, Paul K., and Angela O'Mahony (2013), *A Computational Model of Public Support for Insurgency and Terrorism: A Prototype for More-General Social-Science Modeling*, Santa Monica, Calif.: RAND Corporation, TR-1220-OSD. As of October 13, 2020: http://www.rand.org/pubs/technical_reports/TR1220.html

Davis, Paul K., and Angela O'Mahony (2017), "Representing Qualitative Social Science in Computational Models to Aid Reasoning Under Uncertainty: National Security Examples," *Journal of Defense Modeling and Simulation*, Vol. 14, No. 1, pp. 57–78.

Davis, Paul K., Peter Wilson, Jeongeun Kim, and Junho Park (2016), "Deterrence and Stability for the Korean Peninsula," *Korean Journal of Defense Analyses*, Vol. 28, No. 1, pp. 1–23. As of October 13, 2020: https://www.rand.org/pubs/external_publications/EP66368.html

Dewar, James A. (2002), *Assumption-Based Planning: A Tool for Reducing Avoidable Surprises*, Cambridge, UK: Cambridge University Press.

Dobbs, Michael (1997), *Down with Big Brother: The Fall of the Soviet Empire*, 1st ed., New York: Alfred A. Knopf.

Dobbs, Michael (2008), *One Minute to Midnight: Kennedy, Khrushchev, and Castro on the Brink of Nuclear War*, New York: Alfred A. Knopf.

Doshi, Rush (2019), "Hu's to Blame for China's Foreign Assertiveness?" Brookings Institution webpage. As of October 13, 2020: https://www.brookings.edu/articles/hus-to-blame-for-chinas-foreign-assertiveness

Duelfer, Charles A., and Stephen Benedict Dyson (2011), "Chronic Misperception and International Conflict: The U.S.-Iraq Experience," *International Security*, Vol. 36, No. 1, pp. 73–100.

Ferguson, Neil M., Daniel Laydon, Gemma Nedjati-Gilani, Natsuko Imai, Kylie Ainslie, Marc Baguelin, Sangeeta Bhatia, Adhiratha Boonyasiri, Zulma Cucunubá, Gina Cuomo-Dannenburg, et al., (2020), *Report 9: Impact of Non-Pharmaceutical Interventions (NPIs) to Reduce COVID-19 Mortality and Healthcare Demand*, London: Imperial College. As of October 13, 2020: https://www.imperial.ac.uk/media/imperial-college/medicine/sph/ide/gida -fellowships/Imperial-College-COVID19-NPI-modelling-16-03-2020.pdf

Freedman, Lawrence D., and Jeffrey Michaels (2019), *The Evolution of Nuclear Strategy: New, Updated and Completely Revised*, 4th ed., London: Palgrave Macmillan.

Frum, David, and Richard Perle (2004), *An End to Evil: How to Win the War on Terror*, New York: Ballantine Books.

Gans, John (2019), *White House Warriors: How the National Security Council Transformed the American Way of War*, New York: Liveright.

Garthoff, Raymond L. (1978), "On Estimating and Imputing Intentions," *International Security*, Vol. 2, No. 3, pp. 22–32.

Garthoff, Raymond L. (1988), "Cuban Missile Crisis: The Soviet Story," *Foreign Policy*, Vol. 72, pp. 61–80.

George, Alexander L. (2003), "The Need for Influence Theory and Actor-Specific Behavioral Models of Adversaries," *Comparative Strategy*, Vol. 22, No. 5, pp. 463–487.

George, Alexander L., and William E. Simons, eds., (1994), *The Limits of Coercive Diplomacy*, 1st ed., Boulder, Colo.: Westview Press.

George, Alexander L., and Richard Smoke (1974), *Deterrence in American Foreign Policy: Theory and Practice*, New York: Columbia University Press.

Gigerenzer, Gerd, and Reinhar Selten, eds. (2002), *Bounded Rationality: The Adaptive Toolbox*, Cambridge, Mass.: MIT Press.

Goldberg, Jeffrey (2016), "The Obama Doctrine," *The Atlantic*.

Gong Zheng [龚正] (2020), "How Suleimani's Death Will Change the World [苏莱曼尼之死将如何改变世界]," China Institutes of Contemporary International Relations webpage. As of August 11, 2020: http://www.cicir.ac.cn/NEW/opinion .html?id=1855826b-b35c-49a2-87d7-1c2b18792c56

Groves, David G., Edmundo Molina-Perez, Evan Bloom, and Jordan R. Fischbach (2019), "Robust Decision Making (RDM): Application to Water Planning and Climate Policy," in Vincent A. W. J. Marchau, Warren Walker, Pieter J. T. M. Bloemen, and Steven W. Popper, eds., *Decision Making Under Deep Uncertainty*, Cham, Switzerland: Springer, pp. 135–163.

Halberstam, David (2002), *The Best and The Brightest*, New York: Modern Library.

Hotta, Eri (2013), *Japan 1941: Countdown to Infamy*, Toronto: Alfred A. Knopf.

Howard, Michael (1984), "Men Against Fire: Expectations of War in 1914," *International Security*, Vol. 9, No. 1, pp. 41–57.

Hu Jiping [胡继平] (2018), "Japan's Quest for Strategic Independence Is Not So Easy [日本寻求战略自主没那么容易]," China Institutes of Contemporary International Relations webpage.

Huth, Paul K. (1988), *Extended Deterrence and the Prevention of War*, New Haven, Conn.: Yale University Press.

Jervis, Robert (1976), *Perception and Misperception in International Politics*, Princeton, N.J.: Princeton University Press.

Jervis, Robert (1978), "Cooperation Under the Security Dilemma," *World Politics*, Vol. 30, No. 2, pp. 167–214.

Jervis, Robert (2006a), "Reports, Politics, and Intelligence Failures: The Case of Iraq," *Journal of Strategic Studies*, Vol. 29, No. 1, pp. 3–52.

Jervis, Robert (2006b), "Understanding Beliefs," *Political Psychology*, Vol. 27, No. 5, pp. 641–663.

Jervis, Robert (2017), *How Statesmen Think: The Psychology of International Politics*, Princeton, N.J., and Oxford, UK: Princeton University Press.

Jones, Seth G. (2019), "The Soviet Experience in Afghanistan: Getting History Right," *Lawfare Blog*. As of October 22, 2020: https://www.lawfareblog.com/soviet-experience-afghanistan-getting-history-right

Kahan, James P., William L. Schwabe, and Paul K. Davis (1985), *Characterizing the Temperaments of Red and Blue Agents: Models of Soviet and U.S. Decisionmakers*, Santa Monica, Calif.: RAND Corporation, N-22350. As of October 22, 2020: https://www.rand.org/pubs/notes/N2350.html

Kahneman, Daniel (2011), *Thinking, Fast and Slow*, New York: Farrar, Straus and Giroux.

Kahneman, Daniel, Dan Lovallo, and Olivier Sibony (2019), "A Structured Approach to Strategic Decisions: Reducing Errors in Judgment Requires a Disciplined Process," *MIT Sloan Management Review*, Vol. 60, No. 3, pp. 1–10.

Kahneman, Daniel, and Amos Tversky (1979), "Prospect Theory: An Analysis of Decision Under Risk," *Econometrica*, Vol. 47, No. 2, pp. 263–292.

Klein, Gary A. (1999), *Sources of Power: How People Make Decisions*, Cambridge, Mass.: MIT Press.

Klein, Gary A. (2001), "The Fiction of Optimization," in Gerd Gigerenzer and Reinhard Selten, eds., *Bounded Rationality: The Adaptive Tookit*, Cambridge, Mass.: MIT Press, pp. 103–121.

Lebow, Richard Ned (1983), "The Deterrence Deadlock: Is There a Way Out?" *Political Psychology*, Vol. 4, No. 2, pp. 333–354.

Lempert, Robert J. (2019), "Robust Decision Making (RDM)," in Vincent A. W. J. Marchau, Warren E. Walker, Pieter J. T. M. Bloemen, and Steven W. Popper, eds., *Decision Making Under Deep Uncertainty*, Cham, Switzerland: Springer, pp. 23–51.

Lempert, Robert J., Steven W. Popper, and Steven C. Bankes (2003), *Shaping the Next One Hundred Years: New Methods for Quantitative, Long-Term Policy Analysis*, Santa Monica, Calif.: RAND Corporation, MR-1626-RPC. As of October 22, 2020: http://www.rand.org/pubs/monograph_reports/MR1626.html

Levy, Jack S. (2008), "Deterrence and Coercive Diplomacy: The Contributions of Alexander George," *Political Psychology*, Vol. 29, No. 4, pp. 537–552.

Mandel, Robert (2009), "On Estimating Post–Cold War Enemy Intentions," *Intelligence and National Security*, Vol. 24, No. 2, pp. 194–215.

March, James G., and Johan P. Olsen (2011), "The Logic of Appropriateness," in Robert E. Goodin, ed., *The Oxford Handbook of Political Science*, Oxford, UK: Oxford University Press, pp. 478–497.

Marchau, Vincent A. W. J., Warren E. Walker, Pieter J. T. Bloemen, and Steven W. Popper, eds. (2019), *Decision Making Under Deep Uncertainty: From Theory to Practice*, Cham, Switzerland: Springer.

Mattis, James (2018), *Summary of the 2018 National Defense Strategy of the United States of America: Sharpening the American Military's Competitive Edge*, Washington, D.C.: U.S. Department of Defense. As of October 22, 2020: https://dod.defense.gov/Portals/1/Documents/pubs/2018-National-Defense -Strategy-Summary.pdf

May, Ernest R., and Philip D. Zelikow, eds. (2002), *The Kennedy Tapes: Inside the White House During the Cuban Missile Crisis, concise edition*, New York: W. W. Norton & Co., Inc.

Mazarr, Michael J. (2018), *Understanding Deterrence*, Santa Monica, Calif.: RAND Corporation, PE-295-RC. As of October 22, 2020: https://www.rand.org/pubs/perspectives/PE295.html

Mazarr, Michael J., Arthur Chan, Alyssa Demus, Bryan Frederick, Alireza Nader, Stephanie Pezard, Julia A. Thompson, and Elina Treyger (2018), *What Deters and Why: Exploring Requirements for Effective Deterrence of Interstate Aggression*, Santa Monica Calif.: RAND Corporation, RR-2451-A. As of October 22, 2020: https://www.rand.org/pubs/research_reports/RR2451.html

McFadden, Robert D. (1994), "George W. Ball Dies at 84; Vietnam's Devil's Advocate," *New York Times*, section 1, p. 26.

McMaster, H. R. (1998), *Dereliction of Duty: Johnson, McNamara, the Joint Chiefs of Staff, and the Lies That Led to Vietnam*, Boston: HarperCollins Publishers.

Mearsheimer, John J. (1983), *Conventional Deterrence*, Ithaca, N.Y.: Cornell University Press.

Mearsheimer, John J. (2003), *The Tragedy of Great Power Politics*, New York and London: W. W. Norton & Company.

Merson, Martin (1988), "On the Treadmill to Pearl Harbor," *Journal of Historical Review*, Vol. 8, No. 2, pp. 205–217.

Monbauer, Annika (2014), "The Debate on the Origins of World War One," British Library webpage. As of October 22, 2020: https://www.bl.uk/world-war-one/articles/ the-debate-on-the-origins-of-world-war-one

Morgan, Patrick M. (1983), *Deterrence: A Conceptual Analysis*, 2nd. ed., Cambridge, UK: Cambridge University Press.

Morgan, Patrick M. (2003), *Deterrence Now*, Cambridge, UK: Cambridge University Press.

Murphy, Sean D. (2004), "Assessing the Legality of Invading Iraq," *Georgetown Law Journal*, Vol. 92, No. 4, pp. 173–257.

National Research Council of the National Academies (2009), *Informing Decisions in a Changing Climate*, Washington, D.C.: National Academies Press.

National Research Council of the National Academies (2014), *U.S. Air Force Strategic Deterrence Analytic Capabilities: An Assessment of Tools, Methods, and Approaches for the 21st Century Security Environment*, Washington, D.C.: National Academies Press. As of October 22, 2020: http://www.nap.edu/openbook.php?record_id=18622

Nye, Joseph S., Jr. (2020), *Do Morals Matter? Presidents and Foreign Policy from FDR to Trump*, New York: Oxford University Press.

Office of the Historian (undated), "The 1973 Arab-Israeli War," webpage. As of May 5, 2020: https://history.state.gov/milestones/1969-1976/arab-israeli-war-1973

Pillsbury, Michael (2015), *The Hundred-Year Marathon: China's Secret Strategy to Replace America as the Global Superpower*, New York: Henry Holt and Company.

Post, Jerrold M., ed. (2008), *The Psychological Assessment of Political Leaders*, Ann Arbor, Mich.: University of Michigan Press.

Renshon, Stanley A., ed. (2008), "The Enduring Legacy of Alexander L. George: A Symposium," *Political Psychology*, Vol. 29, No. 4.

Ripsman, Norrin M., and Jack S. Levy (2008), "Wishful Thinking or Buying Time? The Logic of British Appeasement in the 1930s," *International Security*, Vol. 33, No. 2, pp. 148–181.

Ritchey, Tom (2011), *Wicked Problems—Social Messes: Decision Support Modeling with Morphological Analysis*, Berlin: Springer-Verlag Berlin Heidelberg.

el-Sadat, Anwar (1977), *In Search of Identity: An Autobiography*, New York: Harper and Row.

Schachter, Jonathan M. (2002), *The Eye of the Believer: Psychological Influences on Counter-Terrorism Policy-Making*, dissertation, Pardee RAND Graduate School, Santa Monica, Calif.: RAND Corporation, RGSD-166. As of October 22, 2020: http://www.rand.org/pubs/rgs_dissertations/RGSD166.html

Schelling, Thomas (1962), "Foreword," in Roberta Wohlstetter, *Pearl Harbor: Warning and Decision*, Stanford, Calif.: Stanford University Press.

Schrader, Matt (2020), *Friends and Enemies: A Framework for Understanding Chinese Political Interference in Democratic Countries,* Washington. D.C.: Alliance for Security Democracy.

Schultz, Kenneth A. (2001), *Democracy and Coercive Diplomacy*, Cambridge, UK: Cambridge University Press.

Schwenk, Charles R. (1984), "The Use of Devil's Advocates in Strategic Decision-Making," Urbana-Champaign: University of Illinois at Urbana-Champaign, Faculty Working Paper No. 1036. As of April 25, 2020:
https://www.ideals.illinois.edu/bitstream/handle/2142/29170/
useofdevilsadvoc1036schw.pdf?sequence=1

Seeking Truth (2019), "Using Tariffs to Exert Pressure Is Just Another Paper Tiger [用关税极限施压只不过是又一只纸老虎]," Qstheory.cn webpage.
As of August 11, 2020:
http://www.qstheory.cn/international/2019-05/14/c_1124493372.htm

Sheehan, Edward R. F. (1973), "Sadat's War," *New York Times*.

Shih, Gerry (2020), "China Threatens Invasion of Taiwan in New Video Showing Military Might," *Washington Post*.

Simkins, J. D. (2019), "'We'll See How Frightened America Is'—Chinese Admiral Says Sinking U.S. Carriers Key to Dominating South China Sea," *Navy Times*.

Simon, Herbert A. (1957), "A Behavioral Model of Rational Choice," in *Models of Man: Social and Rational: Mathematical Essays on Rational Human Behavior in a Social Setting*, 1st ed., New York: Wiley.

Simon, Herbert A. (1978), "Herbert Simon Prize Lecture: *Rational Decision-Making in Business Organizations*," NobelPrize.org. As of October 22, 2020:
https://www.nobelprize.org/prizes/economic-sciences/1978/simon/lecture/

Smoke, Richard (1977), *War: Controlling Escalation*, Cambridge, Mass.: Harvard University Press.

Stein, Janice Gross (1992), "Deterrence and Compellence in the Gulf, 1990–91: A Failed or Impossible Task?" *International Security*, Vol. 17, No. 2, pp. 147–179.

Stein, Janice Gross (2008), "Crisis Management: Looking Back to Look Forward," *Political Psychology*, Vol. 29, No. 4, pp. 553–569.

Su, Alice (2019), "With Each Generation, The People of Taiwan Feel More Taiwanese — and Less Chinese," *Los Angeles Times*. As of October 22, 2020:
https://www.latimes.com/world/asia/la-fg-taiwan-generation-gap-20190215
-htmlstory.html

Tannenwald, Nina (2007), *The Nuclear Taboo: The United States and the Non-Use of Nuclear Weapons Since 1945*, Cambridge, UK: Cambridge University Press.

Tetlock, Philip E. (2017), *Expert Political Judgment: How Good Is It? How Can We Know?* new ed., Princeton, N.J., and Oxford, UK: Princeton University Press.

Tetlock, Philip E., and Dan Gardner (2015), *Superforecasting: The Art and Science of Prediction*, New York: Crown Publishers.

Tyler, Patrick E. (2004), "Annan Says Iraq War Was 'Illegal,'" *New York Times*.

U.S. Senate (2019), *Taiwan Allies International Protection and Enhancement Initiative (TAIPEI) Act of 2019*, Washington, D.C.: U.S. Government Printing Office. As of October 22, 2020:
https://www.congress.gov/bill/116th-congress/senate-bill/1678/text

U.S. Strategic Command (2006), *Deterrence Operations Joint Operating Concept, Version 2.0*, Washington, D.C.: U.S. Department of Defense.

Walker, Stephen G., and Akan Malici (2011), *U.S. Presidents and Foreign Policy Mistakes*, Stanford, Calif.: Stanford University Press.

Wang Jin [王锦] (2020), "How Should We View Trump's New Middle-East Peace Plan? [如何看待 "中东和平新计划" ?]," China Institutes of Contemporary International Relations webpage. As of August 11, 2020:
http://www.cicir.ac.cn/NEW/opinion
.html?id=04cde127-6e0a-4101-b595-ed4ffb9e50b6

Watson, Cynthia (2019), "The Politics of Humiliation as a Driver in China's View of Strategic Competition," in Nicole Peterson, ed., *Chinese Strategic Intentions: A Deep Dive into China's Worldwide Activities*, Washington, D.C.: U.S. Department of Defense, pp. 17–21. As of April 25, 2020:
https://nsiteam.com/social/wp-content/uploads/2019/10/SMA-Chinese-Strategic-Intentions-White-Paper-FINAL-01-Nov-2.pdf

Weston, Drew, Pavel S. Blagov, Keith Harenski, Clint Kilts, and Stephan Hamann (2006), "Neural Bases of Motivated Reasoning: An fMRI Study of Emotional Constraints on Partisan Political Judgment in the 2004 U.S. Presidential Election," *Journal of Cognitive Neuroscience*, Vol. 18, No. 11, pp. 1947–1958.

Wohlstetter, Albert (1959), "The Delicate Balance of Power," *Foreign Affairs*, Vol. 37, No. 2, pp. 211–234.

Wohlstetter, Roberta (1962), *Pearl Harbor: Warning and Decision*, Stanford, Calif.: Stanford University Press.

Wong, Edward, and Ana Swanson (2020), "Some Trump Officials Take Harder Actions on China During Pandemic," *New York Times*.

Woods, Kevin M., Michael R. Pease, Mark E. Stout, Williamson Murray, and James G. Lacey (2006), *The Iraqi Perspectives Report: Saddam's Senior Leadership on Operation Iraqi Freedom from the Official U.S. Joint Forces Command Report*, Annapolis, Md.: Naval Institute Press.

Woods, Kevin M., and Mark E. Stout (2010), "Saddam's Perceptions and Misperceptions: The Case of 'Desert Storm,'" *Journal of Strategic Studies*, Vol. 33, No. 1, pp. 5–41.

Woodward, Robert (1991), *The Commanders*, New York: Simon & Schuster.

Wuthnow, Joel (2020), "Just Another Paper Tiger? Chinese Perspectives on the U.S. Indo-Pacific Strategy," *Strategic Forum*, No. 305, pp. 3–10.

Yarhi-Milo, Keren (2014), *Knowing the Adversary: Leaders, Intelligence, and Assessment of Intentions in International Relations*, Princeton, N.J.: Princeton University Press.

Zhang, Jian (2014), "China's New Foreign Policy Under Xi Jinping: Towards 'Peaceful Rise 2.0'?" *Global Change, Peace & Security*, Vol. 27, No. 1, pp. 5–19.

Zhang, Suisheng (2013), "China: A Reluctant Global Power in Search of Its Rightful Place," in Vidya Nadkarni and Norma C. Noonan, eds., *Emerging Powers in a Comparative Perspective: The Political and Economic Rise of the BRIC Countries*, New York: Bloomsbury Academic, pp. 101–128.

Zhang Wenzong [张文宗] (2020a), "Has Trump's Impeachment Case Ended? [特朗普弹劾案终结？]," China Institutes of Contemporary International Relations webpage. As of August 11, 2020:
http://www.cicir.ac.cn/NEW/opinion
.html?id=27a2da48-44a8-4ce1-83d1-3bf6f8dc0920

Zhang Wenzong [张文宗] (2020b), "How the U.S. Will Resist the Epidemic in a 'State of Emergency, ["紧急状态"下美国将如何抗疫]," China Institutes of Contemporary International Relations webpage. As of August 11, 2020:
http://www.cicir.ac.cn/NEW/opinion
.html?id=c6de3aad-3fdb-4d4f-9768-acd79aae81ed

Zhao, Suisheng (2004), "Chinese Foreign Policy: Pragmatism and Strategic Behavior," in Suisheng Zhao, ed., *Chinese Foreign Policy: Pragmatism and Strategic Behavior*, New York: Routledge, pp. 3–20.

Zhong ChengFang [鍾辰芳] (2020), "New Survey Data: More than 80% of Taiwanese People Do Not Recognize Themselves as Chinese and Approve of Tsai Ying-Wen's Virus Prevention Performance [新民調: 八成以上台灣人不認同自己是中國人對蔡英文政府防疫能力有信心]," Voice of America Cantonese webpage. As of October 22, 2020:
https://www.voacantonese.com/a/poll-shows-over-80percent-taiwanese
-identify-only-as-taiwnese-and-have-confidence-in-government-handling
-coronavirus-202002226/5306500.html

Lightning Source UK Ltd.
Milton Keynes UK
UKHW020620200721
387431UK00007B/190

9 781977 406521